UNIFYING THE UNIVERSE STUDY GUIDE

Second Edition

$1 + 2 + 2^2 + 2^3 + \ldots + 2^{62} + 2^{63}$?

1.84×10^{19}

H. Padamsee

Cornell University

LINUS Learning

Published by Linus Learning
Ronkonkoma, NY 11779

ISBN 10: 1-60797-607-2

ISBN 13: 978-1-60797-607-3

Printed in the United States of America.

This book is printed on acid-free paper.

Print Number 5 4 3 2 1

TABLE OF CONTENTS

CHAPTER 1

CHAPTER 2

CHAPTER 3

CHAPTER 4

CHAPTER 5

CHAPTER 6

CHAPTER 7

CHAPTER 8

CHAPTER 9

CHAPTER 10

CHAPTER 11

CHAPTER 12

CHAPTER 1

Reading Questions

P. 3: Find one example of astronomic knowledge from each of the following ancient cultures: Babylonian, Egyptian, Jewish, Hindu, Chinese, Native American, Greek.

P. 4: Give your own example of why some would believe that the behavior of nature is chaotic.

Give your own example from motion in the heavens of why some would be led to the conclusion that there is order in natural behavior.

Give your own example of the continuing Greek presence in Western culture.

Pick a Greek god and show how his/her attributes reflect human characteristics.

P. 5: List some the Greek cultural attitudes that proved important to the progress of scientific thought.

Give some of the arguments in favor of flat earth.

P. 7: Mention one example each of how Babylonian and Egyptian culture influenced Greek art and Greek science.

P. 8-9: In what ways was the ancient Egyptian culture superior to the Greek culture at the time of Pythagoras?

In what ways was the ancient Babylonian culture superior to the Greek culture at the time of Pythagoras?

P. 11: Give one example of how Thales advanced upon the knowledge he acquired from the Egyptians.

Give one example of how Pythagoras advanced upon the knowledge he acquired from the Egyptians.

P. 12: Work out the relationship between the height of a tree and the length of its shadow when the angle between the horizontal and the rays of the sun is 30 degrees.

P. 14: Find another example of art that is characteristic of the Greek geometric period.

P. 15: For what puzzling aspects of nature did Thales' cosmology provide rational explanations?

P. 17: Give an example of how the Greek philosophers separated celestial and terrestrial as distinct regions.

How did debate propel the progress of scientific thought after Thales?

P. 18: Explain why Pythagoras thought there is beauty in nature.

What did Pythagoras mean by "nature is mathematical"?

P. 23: What is the relationship between order and symmetry? Give examples.

In what ways did Thales' cosmology lack symmetry?

Are exact symmetries present in nature or only in mathematics? If yes, give an example of an exact symmetry present in nature.

P. 25: What are the arguments against a spherical shaped earth?

What problems can arise by relying purely on aesthetic principles for advancing understanding?

P. 27: Give two examples of geometric relations present in nature.

Read the section on number patterns in this Guide. Identify a number pattern present in nature (not musical harmony).

P. 28: Read the section on alternate proof of Pythagoras theorem. Pick one that uses symmetry arguments.

P. 31: Find an example of an irrational number other than one which can be expressed as a root of a natural number.

P. 34: Pick out some of the arguments in favor of the big bang theory for the creation of the universe.

P. 35: Give examples for some of the natural forces that make up our present world-view.

P. 36: In seeking explanations for natural phenomena, how can we narrow down or select properly from the large number of creative possibilities that would spring from pure imagination.

Give your own example of how senses can trick us into false conclusion.

P. 37: Explain how salt dissolves in water in terms of Democritus' atomic picture.

Explain how perfume spreads in air in terms of Democritus' atomic picture.

P. 38: What properties did Democritus assign to the atoms of the Greek elements?

P. 41: Why did some of Anaxagoras' contemporaries reject his idea that the sun must be fiery rock?

Make a simple flow chart for Socrates' deductive method.

P. 42: Of what crimes was Socrates found guilty?

P. 44: What aspects of the Greek Academy of Athens survive in today's universities?

Why did Plato emphasize the study of geometry at the Academy?

P. 45: Why did Plato conclude that there are five and only five elements?

P. 47: Make a two column comparison list for Plato's world of form and the world of substance.

P. 48: Give some examples of "ideals" in the world of human interactions.

Give some examples of "ideals" in science.

Give some examples of "ideals" in art.

Why did Plato belittle the value of experimental tests?

P. 50-51: Which methods introduced by Aristotle proved important for the long-term progress of science?

P. 53: What differences would you expect in the observations of the appearance of a distant ship from two observers, one located at sea level and the other located at the top of a tall light-house on the same beach?

P. 54-55: List the innovative accomplishments of early Greek thinkers crucial to the progress of science.

P. 56: What do the symmetries of crystalline gems tell us about the structure of matter?

P. 58-60: Besides aesthetics, what are some of the other roles that symmetry plays in the nature and our efforts to understand nature?

P. 60: Give a general definition of invariance symmetry illustrated by one example.

P. 61: What is the relationship between the x,y.z co-ordinate system and its mirror image system (x′, y′, z′) if the mirror is placed parallel to the x-y plane at a certain distance z from the origin? Write down the relationships between x and x′, y and y′, and z and z′.

Is mirror symmetry strictly obeyed in all physical laws? If no, where are violations found?

P. 62: What is the difference between general and special symmetries in physical laws?

What special symmetries are sometimes violated?

What general symmetries are sometimes violated?

Math Based Topics

Number Patterns

Karl Friedrich Gauss was a mathematician and physicist who lived around the year 1800. Even as a child in elementary school, Gauss showed extraordinary talent in mathematics. He irritated his arithmetic teacher by always finishing his assignments far ahead of the end of the period. One day, to occupy his time, the teacher assigned him what he felt would be a sure busy-work problem - to find the sum of the first 100 integers. Gauss was back with the answer in less than a minute. Instead of laboriously summing the integers in sequence, Gauss came up with an elegant trick:

1+ 100 = 101, 2 +99 = 101, ...50 +51 = 101.

With 50 such pairs, the elegant answer was 50 × 101 = 5050.

Another version of the same idea.

Write down the series once forward and once backward.

1 + 2 + 3 + 4 + 5 + ...100

100+ 99 + 98 + 97 + 96 +....1

Add the two rows

101 + 101 + 101 + 101 + 101 ...+ 101..

There are 100 such terms, so the sum of the two rows is 10100

Hence the value of one row is 10100/2 = 5050.

Galileo's discovery about free fall!

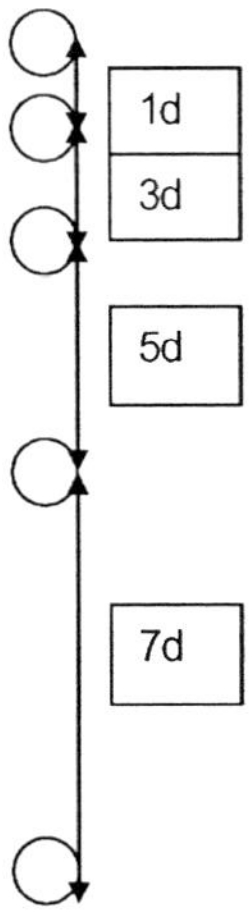

Fig 1.1

Apply the same technique to determine the sum of n odd numbers is n^2.

Pythagoras recognized this principle in a geometric way, see page 32 of the main text.

Where do we see the odd number pattern in nature? In the 1600's Galileo recognized that for equal time intervals, the distance intervals an object in free fall travels increases as the odd numbers, shown above. Therefore he concluded that the total distance the object falls in time t increases at t^2.

Zeno's Paradox

A favorite disciple of Parmenides (Chap 1) and a tutor of Pericles' (Chap 2) , Zeno tried to logically demonstrate the impossibility of motion.

There was no way that the legendary runner could have reached Athens from Marathon because, in order to traverse the distance between Marathon and Athens, he first had to cover half that distance. Then he had to travel half that distance again, and so on, ad infinitum. Taking the distance from Marathon to Athens as unity, the runner must travel the distances of 1/2 + 1/4 + 1/8... Because there is an infinity of terms in this series, the runner can never reach his goal. Motion does not get you anywhere. Motion is an illusion. Only the deceptive senses lead us to believe in the existence of motion.

In full agreement with his master Parmenides, Zeno claimed how logic demonstrates that the senses are illusory. A bit of modern manipulation with infinite series shows how the fallacy in Zeno's paradox stems from his intuitive notion that the sum of an infinite series of numbers should be infinite. This is wrong. It can be rigorously proved that the *infinite* sum 1/2 + 1/4 + 1/8 + 1/16 + 1/32 +... is equal to the *finite* number 1. The runner will indeed reach his goal. Motion is not an illusion.

Let $A=\frac{1}{2}+\frac{1}{4}+\frac{1}{8}+\frac{1}{16}+.......$ We wish to know the value of A,

Then $2A = 2\times\frac{1}{2}+2\times\frac{1}{4}+2\times\frac{1}{8}+2\times\frac{1}{16}+........$

Multiplying out, $2A = 1+\frac{1}{2}+\frac{1}{4}+\frac{1}{8}+\frac{1}{16}+........$

Now Subtract the first line from the third line

$$2A\text{-}A = 1+(\frac{1}{2}+\frac{1}{4}+\frac{1}{8}+\frac{1}{16}+........)$$

$$-(\frac{1}{2}+\frac{1}{4}+\frac{1}{8}+\frac{1}{16}+........) = 1$$

The two infinite series A in the parantheses cancel each other out

So 2A - A =1, A = 1

The Story of Chess

One story claims that the game of Chess is a variant of Chaturanga, a game that was invented in India. Others claim it was invented in China, but the story about the Indian invention is a fascinating one.

A young king who lived in India was bored of the normal royal amusements, so he asked his minister to come up with a new game; one that would be much more challenging. After several tries the minister came up with Chaturanga based on pieces moving on an 8 × 8 square board with black and white squares interspersed, like the modern Chess board. Pieces captured the opponent's pieces as in a familiar battle. The king was fascinated by the game and wished to reward the inventor with a fine prize. "Name anything in my kingdom that you would like," he magnanimously offered. "I would like one grain of rice for the first square, two grains for the second, four grains for the third and so on doubled for each of the 64 squares of the game board." "Why, that is too easy, exclaimed the king? Are you sure you would rather not have my best gold or diamond jewel?" "No," said the minister, "the rice will be more than enough."

When the king ordered the rice from his granaries, he soon learned that all would be exhausted even before they reached the 40th square. Not all the riches in the entire kingdom would be sufficient to meet the reward. Whereupon the minister magnanimously offered to relieve the king of one of his beautiful wives instead !

Fibonacci Sequence

In the year 1202, Italian mathematician Fibonacci investigated a population problem that led to the now famous Fibonacci series. How fast rabbits breed in the following simple, theoretical situation? Suppose we start with one male and one female.

Rabbits can mate at the age of one month so that at the end of its second month a female can produce another pair of rabbits. The female always produces one new pair (one male, one female) every month from the second month on. Fibonacci solved the population growth pattern ignoring the death of rabbits.

1. At the end of the first month, they mate, but there is still one only 1 pair.
2. At the end of the second month the female produces a new pair, so now there are 2 pairs of rabbits in the field.
3. At the end of the third month, the original female produces a second pair, making 3 pairs in all in the field.
4. At the end of the fourth month, the original female has produced yet another new pair, the female born two months ago produces her first pair also, making 5 pairs.

   The number of pairs of rabbits in the field at the start of each month is 1, 1, 2, 3, 5, 8, 13, 21, 34, ...

Chapter Questions

1. Give an example of creative thinking by the Greeks to come with a natural explanation.
2. Find an example on your campus of Greek cultural roots. Explain their Greek origin.
3. How did Greek thinkers and artists emphasize the importance of the individual?

   Give examples.
4. Give some specific examples to delineate the difference between a mathematical development and a scientific development.
5. Give one example each of the following symmetries found in (a) mathematics and (b) in nature: translation, rotation, reflection.
6. Give an example of a supernatural explanation of a natural phenomena. Offer a rational explanation for the same.
7. Why would there be any reason to debate the shape of the earth - flat versus spherical? Give rational arguments on both sides. Discuss the resolution.
8. Discuss the difference between Pythagoras' and Aristotle's approach to show that the world is a sphere.
9. Give an example of how debate was important for the progress of science.
10. If the earth is round, why does it appear flat to the casual observer? Think of the size of the earth and the size of the observer. Draw a diagram or two to explain your answer.
11. Discuss one connection between (a) science and art (b) science and music.
12. Give two arguments to support the idea that there must be an order underlying the seeming chaos in the behavior of nature.
13. Discuss the natural and rational aspects of Empedocles' cosmogony.
14. Discuss the influence of Plato's ideas about "form and substance" over Aristotle's bifurcation of the cosmos.
15. Give an example of an observation that can be deceiving.
16. For the development of science, what was the *key difference for science* between the Egyptian treatment of mathematics and the Greek treatment?
17. Why did Thales think of the earth as a disk?
18. Why did Thales think of the earth-disk as *floating* on water?

19. Why did Thales' idea that water may be a universal element become a lasting contribution to the development of science.

20. One of the most notable quotes from Pythagoras is:

    "All is number" What did he mean by this?

21. Why did Pythagoras conclude that the earth must be a sphere and not a disk floating on water.

22. What objections did some of Pythagoras contemporary thinkers raise against the idea of a spherical earth?

    "Take but degree away, untune that string,

    And hark, what discord follows. Each thing meets

    In mere oppugnancy. The bounded waters

    Should lift their bosoms higher than the shores

    And make a sop of all this solid globe."

23. In the above verse, whose ideas does Shakespeare refer to?

24. Give two examples that could have motivated Pythagoras to believe that the operations of nature are governed by principles that reflect beauty?

Math Based Problems

1. Find another Pythagorean triplet which is not a scaled replica of the 3,4,5 triplet.

2. Knowing that the area of a circle is πr^2, convince yourself that Pythagoras' theorem still applies if squares are replaced by semi-circles.

3. Using the methods discussed in this book on number patterns show that the sum of the first n odd numbers is n^2. Where does the above number pattern show up in nature?

4. Using a symmetry argument, show with diagrams how the formula for determining the area of a right angle triangle can be derived from the formula for determining the area of the rectangle.

5. By how many degrees would you have to rotate an equilateral triangle (all angles are equal) so that it would remain the same. Same question for a square.

6. Which of the following triangles do not have a right angle? When the sides are:

a.	3, 4, 5	b.	12, 16, 20
c.	9, 12, 16	d.	6, 8, 10
e.	7, 24, 25	f.	8, 15, 17.

7. Use algebra to show that there is only one right triangle whose sides are consecutive natural (positive) numbers.

8. Draw a right angle triangle whose hypotenuse is $\sqrt{6}$. Draw diagrams to show the steps that lead to the final triangle.

9. If the first odd number is 1, the second is 3, the third is 5, what is the nth odd number in terms of n?

10. Using Gauss' trick, prove Pythagoras' discovery that the sum of the first n odd numbers = n^2.

11. Using symmetry arguments prove (a) that the angles at the base of an isosceles triangle are equal and (b) that vertically opposite angles are equal.

12. What is the sum of the first 1000 numbers?

13. Write down the Fibonacci series of numbers starting with 1, 2.... Determine the ratios of successive pairs, e.g. 2/1 = 2... After determining about TEN ratios, find the pattern in the value of the ratios, and guess the

closest answer for what the ratio will become for the 100 th pair (there is no need to work with 100 pairs!). Plot the ratios on the y-axis, the x-axis being the first, second, third ratios and so on.

14. Consider a rectangle with sides short side a and long side b. A golden rectangle (defined by Greek architects) is one for which the ratio $b/a = (b+a)/b$. Show that the ratio b/a is same as the answer to Question 13 above.

15. If a cold virus molecule regenerates into two molecules every minute, approximately how many viral molecules will fill the room in one hour after the first virus enters the room?

16. Will you accept the following bet? If you can withstand the weight of one million one-dollar bills on your belly for 10 minutes, you can have the million dollars. Explain your answer using qualitative reasoning and number estimates.

17. In the story about the invention of CHESS, calculate how many grains of rice would be needed for the last square.

    Estimate how many grains of rice are there in a 20 lb bag which costs about $10.

    Estimate the cost of populating the last square, compare that with the US GNP.

    Estimate the cost of populating all the squares.

Exploration Topics

1. Explore on your own (using the web) some of the other ideas about the cosmos and the elements advanced by Greek philosophers who came after Thales. Give their names and their theories.

2. Look up the vibration frequency (cylces/sec) of the following musical notes :A, C,E, G, A#. Make a table of the ratio of the numbers to see the simple ratios. Play the notes on a piano or other musical instrument to experience the harmony and dischord.

3 Look for a Google-earth photo of earth clearly showing the curvature.

4. In the context of the art presented in this chapter, present a short discussion of the following quote from Robert Hughes:

   "A truly significant work of art is one that prepares the future."

5. Discuss the differences in the attitude towards knowledge and understanding between the Egyptian and the Greek cultures explaining why such differences may have existed. Feel free to draw on material from other sources or courses.

6. Discuss the central ideas about the shape, position and motion of the earth from the passage below. What are the theories? What are the observations? Where are the fallacies?

   From Aristotle's, On the Heavens[7]:

   "The natural motion of the earth as a whole, like that of its parts, is towards the center of the Universe: that is the reason why it is now lying at the center. It might be asked, since the center of both is the same point, in which capacity the natural motion of heavy bodies, or parts of the earth, is directed towards it; whether as center of the Universe or of the earth. But it must be towards the center of the Universe that they move...It so happens that the earth and Universe have the same center, for the heavy bodies move also towards the center of the earth, yet only incidentally, because it has its center at the center of the Universe...

   From these considerations it is clear that the earth does not move, neither does it lie anywhere but at the center. In addition the reason for its immobility is clear from our discussions. It is inherent in the nature of the earth to move from all sides to the center (as observation shows), and of fire to move away from the center towards the extremity, it is impossible for any portion of earth to move from the center except under constraint... If then any portion is incapable of moving from the center, it is clear that the earth itself as a whole is still more incapable, since it is natural for the whole to be in the place towards which the part has a natural motion.

7. Why did Plato accept reason, but reject observation of the real world? Discuss the ramifications of his philosophical stance on the progress of science.

8. The theories of the ancient Greek thinkers have been overthrown or supplanted by modern ones. Does this mean that the ancients contributed nothing to science? Discuss your answer.

9. Discuss the significance of idealization using examples from mathematics, science, art and human behavior.

10. Explore the grand topic of Fibonacci numbers and patterns in nature and mathematics.

    http://www.mcs.surrey.ac.uk/Personal/R.Knott/Fibonacci/

11. Explore the fascinating topics of Pythagoras theorem and advances.

    http://www.mcs.surrey.ac.uk/Personal/R.Knott/Pythag/pythag.html

    See 81 separate proofs of Pythagoras' theorem at http://www.cut-the-knot .org/pythagoras/

Sample Quiz Questions

1. In the figure below the "corresponding angles" A and B between the two parallel lines are equal, < A = < B. What symmetry operation would provide the most elegant proof of this basic theorem in geometry?

   a. Rotational symmetry

   b. Reflection symmetry

   c. Translation symmetry

   d. Bilateral symmetry

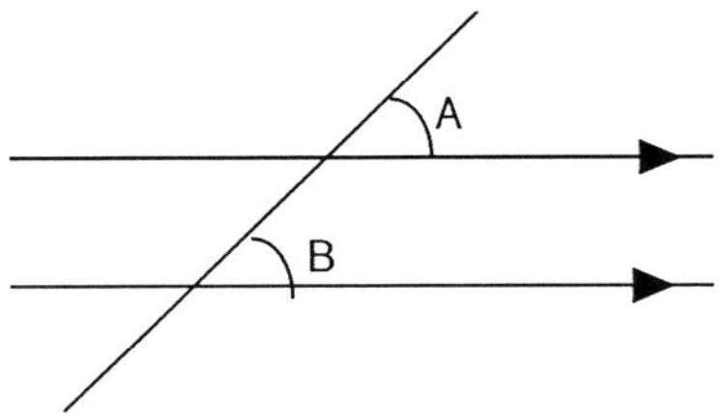

2. What is the sum of the first 1000 numbers?

   a. 5050

   b. 10,000

   c. 50500

   d. 500500

   e. 100,000

3. If a cold virus molecule regenerates into two molecules every minute, approximately how many viral molecules will fill the room in one hour after the first virus enters the room?

   a. one thousand

   b. one million

   c. one billion

   d. one trillion

   e. one billion billion

4. Determine the sum of the first 75 odd numbers

   1 + 3 + 5 + ...= ?

   a. 7500

   b. 5625

   c. 150

   d. 1444

   e. none of the above

5. What aspect(s) of Thales cosmology below were dissatisfying to other Greek thinkers?

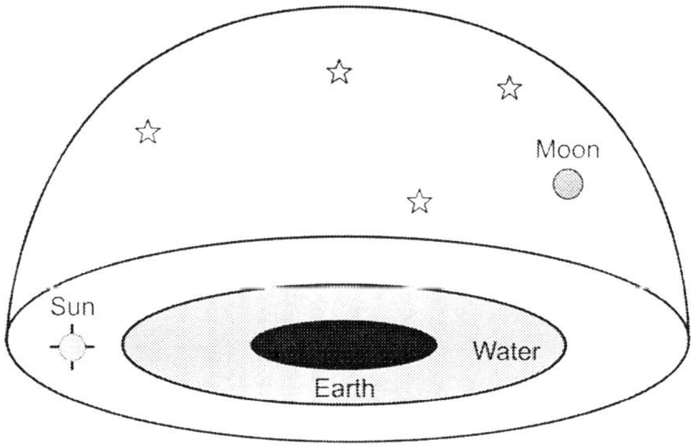

a. The asymmetry of the heavens

b. The flat earth

c. Water is the only element

d. All of the above

6. What is the special property about the north star that other stars do not have?

a. It is visible in the north.

b. It is part of a famous constellation, the little Dipper.

c. It is the only star which does not move during the course of a night.

d. It can only be seen in the northern hemisphere.

7. By how many degrees would you have to rotate an equilateral triangle (all angles are equal) so that it would remain the same?

a. 30 c. 120

b. 60 d. 90

8. The heavens themselves,
The planets and this center,
Observe degree, priority, and place,
Insisture, course, proportion, season, form,
Office, and custom, in all line of order. "
What aspect of nature does Shakespeare refer to in the above?

a. The human mind is capable of discovering nature's laws.

b. The universe is fundamentally orderly.

c. The laws that operate on earth are the same as those which operate in the heavens.

d. Different laws operate on the earth and in the heavens.

9. Determine the height of the tree below

a. 20 meter

b. 10 meters

c. 5 meter

d. none of the above

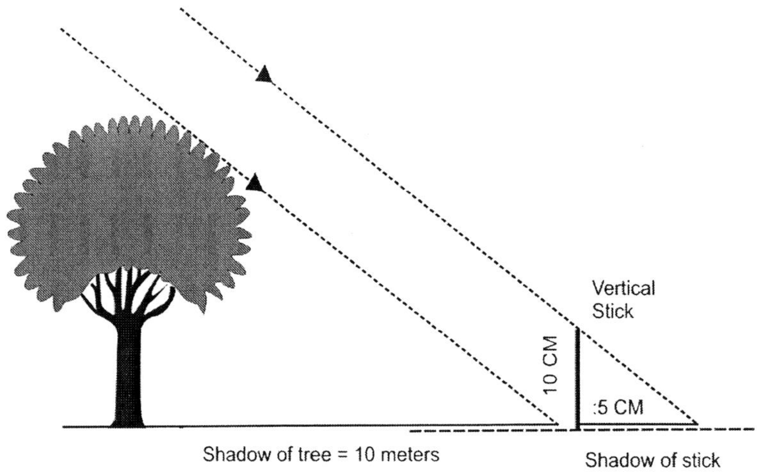

Answers to Chapter 1 Sample Quiz

1. c
2. d
3. e
4. b
5. d
6. c
7. c
8. b

Homework Questions for Chapter 1 (with sample answers)

Your Professor will choose only some of the questions below. Sample answers are provided to some of the questions for students to emulate.

General Directions for all Homeworks.

Work together in Teams to answer questions and take notes in your sessions. If you just split the questions between team members (instead of working together) you will find it difficult to give good answers in section and chapter quizzes. Team work will be very important to arrive at solutions as different members will suggest different creative approaches.

1. Give examples of supernatural explanations of lightning and earthquakes. Offer rational and natural explanations for the same.

   Solution

   In Greek mythology many natural events were ascribed to the actions of the gods. For example, Zeus was responsible for lightning and Poseidon for earthquakes, with little understanding of the structure of the earth, or the origin of electricity, early people had to create supernatural explanations like this to explain the world around them. While it may surprise some of you to know that lightning still isn't fully understood, we do know that it is caused by a build-up of static electricity in the clouds that is then released when an enormous current connects the sky and ground. The bright arc and loud sound are a result of the immense heat that is released by running a current through the normally nonconducting air. Earthquakes are also now understood to be caused by motions of the earth's tectonic plates.

2. Debate the shape of the earth - flat versus spherical by giving rational arguments on both sides. Discuss the resolution.

3. Discuss the difference between Pythagoras' and Aristotle's approach to show that the world is a sphere.

4. Using a symmetry argument, show with diagrams how the formula for determining the area of a right angle triangle can be derived from the formula for determining the area of the rectangle.

5. By how many degrees would you have to rotate an equilateral triangle (all angles are equal) so that it would remain the same. Same question for a square.

6. What is the sum of the first 1000 numbers?

   a. 5050 b. 10,000 c. 50500
   d. 500500 e. 100,000

7. Read about Fibonacci numbers in this Study Guide. Take the ratio of two successive numbers in Fibonacci's series, (1, 1, 2, 3, 5, 8, 13, ..) and divide each by the number before it, to find the following series of numbers: $^1/_1 = 1$, $^2/_1 = 2$, $^3/_2 = 1{\cdot}5$,....Plot the ratios on a graph where the x-axis is the number of the ratio and y-axis is the ratio. Find the value which the ratios seem to be settling down to. Use a calculator to obtain the converging value to 4 decimal places. Take the ratios the other way round i.e. divide each number by the one *following* it: 1/1, 1/2, 2/3, 3/5, 5/8, 8/13, ..? Use your calculator and plot a graph of these ratios and see if anything similar is happening compared with the graph above.

   Using the discussion following Zeno's Paradox, show that the sum of the infinite series: 1/3 + 1/9 + 1/27 +... = ½

8. What do the symmetries of crystalline gems tell us about the structure of matter?

   Numerical estimate questions build on the power of reasoned analysis, and the power of numbers. At first the questions may appear unanswerable, but creative thought opens the door to solutions. Although the question itself is not of great value the method of thought is. Team work will be very important to arrive at solutions as different members will suggest different creative approaches.

9. Will you accept the following bet? If you can withstand the weight of one million one-dollar bills on your belly for 10 minutes, you can have the million dollars. Explain your answer using qualitative reasoning and number estimates. See note below:

10. (Bonus) In the story about the invention of CHESS, calculate how many grains of rice would be needed for the last square. Here are the steps to follow.

    Estimate how many grains of rice there are in a 20 lb bag which costs about $10. (Make your best guess about how much one grain weighs.)

    Estimate the cost of populating the last square, compare that with the US GNP.

    Estimate the cost of populating all the squares.

    Partial Solution

    Make your best guess about how much one grain weighs. Guess 1 gram

    Estimate how many grains of rice there are in a 20 lb bag which costs about $10.

    20 lb = 20 × 454 grams = 10,000 grains approx.

    Estimate the cost of populating the last square, compare that with the US GNP.

    No of grains in the last square is $2^{63} = 10^{19}$ (approx) grains

Divide by 10,000 for number of grains in one bag = 10^{15} bags at \$10 per bag.

Cost is \$ $10^{16} = 10^4 \times 10^{12} = 10^4$ trillion dollars

US GNP today is only 16 trillion.. oops not enough \$ to buy the rice to fill the last square!

Estimate the cost of populating all the squares.

11. The radius of the earth is about 100 times smaller than the radius of the sun. If the angle between the two sun's rays (shown below) to the surface of the earth is 0.25 degrees, what is the distance between the earth and sun, if the radius of the earth is 6370 km.

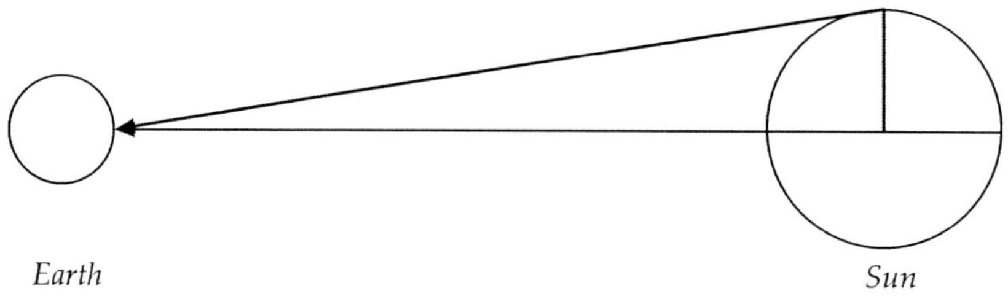

CHAPTER 2

Reading Questions

P. 64: Why is it important to find order in nature?

P. 65: In what way(s) is our search for the fundamental elements connected to the universe?

What role do mathematical patterns play in our quest to understand the workings of the universe?

What is the value of studying the mistakes of Aristotle and others?

P. 67: How did Aristotle introduce order into the cosmos?

Why did Aristotle treat heaven and earth as regions with different substances?

p. 69: Explain why Aristotle argued that a heavy body should fall to the earth faster than a lighter body.

How did Aristotle's theory of motion resolve the problems which arose with the spherical shape of the earth?

According to Aristotle's theory why does the earth need no support?

According to Aristotle's theory why does the earth have a spherical shape?

P. 70-71: How did Aristotle's theory explain the phenomena of lightning, volcanoes and geysers?

According to Aristotle, what is the difference between natural motion and violent motion?

Why was Aristotle's theory so persuasive to last two thousand years?

P. 72: How did Aristotle's followers explain the dark markings on the heavenly body, the moon?

Why did Aristotle explain comets as atmospheric phenomena?

P. 73: Where did Aristotle go wrong in his methodology?

What lessons do we learn from Aristotle's errors?

Why are modern scientific theories not immune to the mistakes of Aristotle?

P. 76: What were some of the factors which contributed to the slow decline of Athens as the dominant seat of cultural advances?

What were some of the factors that contributed to the rise of Alexandria as the new seat of cultural advances?

P. 77: What was the difference between the attitude of Archimedes and that of Plato and Euclid for studying the natural world?

Why does π appear so frequently in mathematical patterns of nature?

P. 81: Why does a helium filled balloon rise through air?

P. 84: Is the earth exactly spherically symmetric if you ignore the mountains and valleys? What is the cause of this asymmetry?

P. 89: Why was there little change in the theory of elements from the early Greeks till the time of Lavoisier?

P. 90: What is the difference between phlogiston and oxygen?

P. 91: What is the nature of calx?

What is needed for a new discovery to have a significant impact on the progress of science?

P. 92: How did Lavoisier show that water and earth are not fundamental elements?

P. 93: Are the elements of the Periodic Table really elementary?

P. 94: What were the differences between Dalton's and Democritus' atomic theories?

P. 95: How can the solution of salt (or sugar) into water be explained by the atomic theory?

How can the atomic theory explain the spreading of perfume's aroma in a room ?

P. 97: Why did Dalton conclude that water is HO, a compound of one atom of H and one atom of O, rather than our familiar formula H_2O.

P. 98: What is the typical voltage of a car battery? Is it DC or AC?

What is the typical voltage across the two terminals of the wall plug? Is it AC or DC?

P. 99: How were the elements sodium and potassium first isolated?

P. 100:What is the nature of a chemical bond between two atoms?

What is the difference between an atom and an ion?

P. 101:Is electric charge discrete or continuous?

How big is the smallest unit of electrical charge?

P. 102:What is the common chemical property of the elements in the first (last) column of the Periodic Table?

How many elements are there in the first, second, third, fourth and fifth rows of the Periodic Table

P. 103:What is the next term in the number series: 2, 8, 18, 32?

P. 104:Where did the elements in the Periodic Table come from?

P. 106:Why does the gas in a neon tube light up?

How did Thomson's discovery change the picture of Dalton's atoms?

P. 107:Why does the positive charge in the atom carry most of its mass?

P. 108:Why did Roentgen choose the name X-rays for his discovery?

P. 110:How do the cathode rays (electrons) become X-rays?

P. 111:Why did the Curies need tons of the pitchblende to extract radium?

P. 112:How did Rutherford determine that alpha ray particles are related to the atoms of helium?

Is it possible for the atom of one element to turn into an atom of another element naturally?

Is it possible for the atom of one element to turn into an atom of another element by artificial means?

P. 113:Why did Rutherford choose gold as the target for bombardment by alpha particles to study the structure of an atom?

P. 114:Why did most of the alpha rays go straight through the thin gold foil with minor deflection?

P. 115:How many electrons are there in the element of atomic number 15?

P. 116:What would be the converse of the statement: When an energized atom drops from a high to a lower energy level, it releases a discrete amount of energy in the form of light.

If the nucleus carries the positive charge and the electrons carry the negative charge in the atom, what prevents the electrons from coalescing with the nucleus?

P. 120:What are the satisfying (dis-satisfying) aspects of Bohr's atomic theory?

P. 122:Why do all the electrons in an iron atom not settle together into the lowest energy state?

P. 123:Why are H, Li and Na very chemically active?

Why are the noble gases so chemically inert?

P. 124:If the atomic weight of He is 4 times the atomic weight of H, why did Rutherford assign only 2 protons to the nucleus of the He atom?

How many protons are there in the element of atomic number 15?

If the protons in the nucleus of element 15 are all positively charged, what prevents them from flying apart?

Why does the element radium emit alpha particles?

P. 125:Why does the neutron fail to make tracks in a cloud chamber as does the proton.

If the nucleus of an "element" has one proton and two neutrons, which element of the periodic table would its chemical properties follow?

P. 126:How did Chadwick identify the presence of neutrons in his cloud chamber without any tracks?

P. 127:Why did Pauli hypothesize the presence of the neutrino in the decay of a neutron?

P. 128:Why does the neutrino interact so weakly with matter?

Besides the weak force, with what other force does the neutrino interact with matter?

Describe what happens when an electron collides with a positron.

P. 129:Besides particle accelerator beams, what is another source for high energy protons to produce new particles?

P. 130:List the major differences in the properties of baryons, mesons and leptons.

P. 133:How did Gell-mann's theory of quarks violate Faraday and Thomson's discoveries about the nature of electric charge?

P. 134:List the "atomic weights" of the 6 quarks relative to the H atom.

Why is the name lepton (light particle) a misnomer for the general family?

P. 136:Give one example of how precision in measurement led to a new discovery about nature.

What are some of the applications of Rutherford's discovery of the nucleus (other than nuclear power plants)?

P. 137:Give an example that supports Einstein's statement that "pure thought can grasp reality."

Give an example supporting Einstein's second quote on the same page.

Math Based Topics

The History of Pi (π)

The number Pi (π), is defined as the ratio of circumference of a circle to its diameter. It is one of the most important constants found in physics and engineering.

There are many interesting stories about its value which cannot be determined exactly as a decimal because π is an irrational number. This means that the value of π can never be expressed as a ratio of two whole numbers, which is equivalent to the statement that its decimal representation is unending and unrepeating. A repeating decimal representation, such as 0.151515...... is really just the ratio of two whole numbers, in this case: 1/66. A supercomputer value of π is known to a trillion decimal places, but is still inexact ! The value of π to 11 decimal places is enough to represent the circumference of the earth to one mm accuracy. So the practical uses of such large number of decimal places is not so important. The fact that π is indeed irrational was mathematically proved by Lambert in 1761.

Thousands of years ago, Egyptian, Babylonian and Indian civilizations had (different) estimates for the use of the value of π in practical perimeter and area problems involving circles. The Egyptians used $(4/3)^4$, the Babylonians used 3 1/8, and the Indians used the square root of 10. All these approximations had an error only of less than 1%, which is quite impressive.

The Bible (which came later) has a less accurate estimate. From Kings chapter 7 verse 23,

> "... and he made a molten sea, ten cubits from one brim to the other: it was round about ... and a line of thirty cubits did compass it round about."

Thus the value of π in the Bible was approximately 3, much less accurate than the older cultures.

Archimedes of Alexandria took the first major step to develop a systematic approach to a more accurate value of π around 250 BCE. He developed a method discussed in lecture using inscribed and circumscribed polygons with 6-, 12-, 48-, 96-sides for calculating better and better approximations to the value of π, to find that $3\ 10/71 < \pi < 3\ 10/70$. Today we often use the latter value 22/7 for work which does not require great accuracy. Later, in about 150 AD, Ptolemy of Alexandria came up with 377/120 and in about 500 AD the Chinese Tsu Ch'ung-Chi gave the value as 355/113. These estimates are correct to 3 and 6 decimal places respectively.

In 1882, Lindemann proved that π is not only irrational, it is also transcendental — that is, it is not the solution of any polynomial equation with integral coefficients. $Ax^3 + Bx^2 + Cx + D = 0$ is an example of such a simple polynomial equation, with coefficients A, B, C and D as integers. Such an equation could have solutions which can be expressed as roots. Another version of this statement is that πr^2 can never be expressed as x^2 for any value of r or x. This means that you can never find a square (side x) with exactly the same area as a circle (radius r). This leads to an old adage: "you can never square a circle."

Many mathematicians are interested in finding the value of π to as many places as possible and on finding expressions or infinite series for π and its approximations, such as those found by the Indian mathematician Ramanujan, shown in Lecture.

You should try and find the value of π to 10 decimal places using one of these expressions. Ramanujan also came up with approximations for π using roots, such as this one:

> 63(17+15root5)/25(7+15root5) = 3.141592654... good to 9 decimal places.

It may be interesting to do some research on the web on other interesting aspects about the history and the mathematics of π, as well as other transcendental numbers, to come up with some interesting facts to discuss.

Avogadro's Number

Before becoming the first physics professor in Italy, Avogadro was a lawyer and mathematician. He knew that chemically active gases such as hydrogen, nitrogen and oxygen are made up of molecules of two atoms each, compared with inert helium gas which is always monatomic. Therefore he thought in terms of their molecular weight instead of their atomic weight. For example the atomic weight of O is 16, but the molecular weight of O_2 is 32. Avogadro made an amazing discovery about the relationship between the volume and molecular weight of such gases. One molecular weight of hydrogen in grams (commonly called one mole) occupies a volume 22.4 liters at room temperature and atmospheric pressure. So also, one molecular weight of nitrogen in grams (28 grams) occupies the same volume of 22.4 liters at room temperature and atmospheric pressure. This is known as Avogadro's Law. Avogadro guessed that the reason this must be true is that 22.4 liters of any gas or one mole of any gas must contain the same number of molecules of that gas ! A marvelous deduction. One can generalize this deduction to one mole of any molecule, independent of its state, whether gas, liquid or solid.

But of course he did not know that number of molecules in a mole. Today it is possible to determine this number in many different ways. For example, x-ray analysis of the cubic salt crystal (NaCl) reveals the structure of the crystal, and the length of the sides of the cube (5.64×10^{-10}) in meters. From that length we determine the volume in cubic meters of one crystal of NaCl is $(1.8 \times 10^{-28}) m^3$. Hence one m^3 will have 5.55×10^{27} unit crystals. X-ray studies also tell that there are 4 molecules of NaCl in one cubic crystal lattice. Therefore there are 2.22×10^{28} molecules of NaCl in one m^3. The density of NaCl is 2.2 gm/cm^3 or 2.2×10^6 gm/m^3. Therefore 2.22×10^{28} molecules have a mass of 2.2×10^6 gm. Since the molecular weight of NaCl is 58.4 (adding together the atomic weights of Na and Cl) we are interested in finding out how many molecules of NaCl there are in just 58.4 gm. This will be Avogadro's number, N_A.

$$N_A = 58.4 \times 2.22 \times 10^{28} / 2.2 \times 10^6 \sim 6 \times 10^{23}$$

Chapter Questions

1. What kind of eclipse do you get when the earth comes between the moon and the sun? lunar or solar?
2. Why did Plato emphasize the ideal world over the real world?
3. Why did Aristotle emphasize the value of observations?
4. What were Aristotle's observations to support the notion that the earth is a sphere?
5. What was Aristotle's theoretical explanation that the earth must be in the shape of a sphere?
6. State five items of evidence from Aristotle and elsewhere that the world is round, not flat.
7. What problems did the idea of a spherical earth bring up? How did Aristotle resolve those problems?
8. If the earth is round, why does it appear flat to the casual observer?
9. Why did Aristotle contend that the laws of nature operating in the heavens must be different from the laws operating on earth?
10. How might Aristotle explain the eruption of a volcano in terms of his understanding of the elements of nature?
11. What was the lasting feature of Aristotle's theory of elements?
12. Give an example (other than the flat earth) of an observation that can be deceiving.
13. Explain how a ship made of iron can float in water even though the density of iron is much larger than the density of water.
14. What was the lasting feature of Aristotle's theory of elements?
15. Why was library of Alexandria important to the progress of science?
16. What asymmetry did Eratosthenes run into in the records of the library?
17. Give three different assumptions Eratosthenes made in his derivation of the size of the earth.
18. What role did abstract geometry play in the measurement of the size of the earth?
19. How did Eratosthenes' interest in details - such as shadow lengths at various locations - show up in other parts of Alexandrian culture in his time or earlier?
20. How do we know that water is not an element?

21. How did Dalton estimate the atomic weight of oxygen relative to hydrogen? Why was his result wrong by a factor 2 or more?

22. Why is the periodic table called periodic?

23. Giving two specific examples from the quest for the chemical elements, discuss why close attention to minor details can be important to the progress of science.

24. Which chemical element is the simplest in its sub-structure? Explain why?

25. What evidence led Faraday to propose the idea that all charges are integral multiples of a single fundamental unit of charge, and that one never finds fractions of that charge?

26. Where do the elements of the Periodic Table come from?

27. What does the fact that there are more than 100 chemical elements tell us about the atoms of the elements?

28. Find the next number in the series 2, 8, 18, ...

29. How many electrons are found in the outermost shell of aluminum (Al, atomic number 13).

30. How do we know that light can be treated as a wave?

31. Is it possible to see atoms with instruments available today?

32. According to a frequently-told story, when someone asked Ernest Rutherford how he always managed to be on top of the wave of progress in physics, he replied: "because I *make* the wave!" Discuss the validity of this claim.

33. Describe under what circumstances is it possible to transform one chemical element of the periodic table into another.

34. When radium (atomic number 88) emits an alpha particle, what is the atomic number and the atomic weight of the resulting nucleus.

35. In the Bohr model of the H atom, what is the relationship between the radii of the permitted electron orbits ?

36. How was the Bohr model similar to the model of the solar system? How was it different? Why was it (temporarily) accepted despite its obvious defects?

37. Arrange the following in order of increasing mass: down-quark, proton, electron, neutrino, H-atom, water molecule.

38. Explain the difference between a proton and a neutron in terms of Gell-mann's theory of quarks.

Math Based Questions

1. Are the rays from the sun arriving at two points on earth exactly parallel? If not, why not? Draw a sketch of the earth and sun and rays.

2. Redraw Eratosthenes' diagram to compare the shadow made by an obelisk at the equator with the shadows made at Syenne and Alexandria.

3. Taking the density of water to be 1 gm/cc, what is the loss of weight in water in gms for an object of density 8 gm/cc with a volume of 10 cc?

4. There are 9 identical-looking stones, but one of these is a valuable gem. All the stones weigh exactly the same, except the gem weighs slightly more. How would you identify the gem by using only 2 attempts with a scale balance?

5. There are two conical shaped glasses, one measures 6 oz and the other 10 oz. How would you fill the larger glass with exactly 8 oz ?

6. A pizza parlor tells you they have 3 size pizzas, small, medium and large. The diameter of the medium is twice the diameter of the small, and the diameter or the large is three times the diameter of the small. The price of the large is the same as the sum of price of the small plus the medium pizzas. What should you order if your goal is to eat as much pizza as possible: one large pizza OR one small and one medium pizza?

7. Estimate the area of the surface of the earth in square meters. From the density of air and the approximate height of the atmosphere, determine the total mass of the atmosphere. Now determine the ratio of mass to area. Convert this number to pounds/sq inch, normally used for atmospheric pressure.

   Use the density of air as constant (1.2 kg m^{-3}) and the atmosphere to terminate abruptly at 7.8 km. Actually the density drops with height so the overall height of the atmosphere is about 10 times higher with no abrupt boundary.

8. Estimate the volume of the earth in cubic meters.

9. With each breath we take we inhale many molecules of the same air which Einstein breathed during his life ! Estimate how many such molecules of Einstein's breaths we inhale in each breath. Follow the many steps outlined below to solve this problem.

   Use the following assumptions and be careful with units.

   E.g. one pound is 454 grams.

   All the air which Einstein breathed in his entire life is by now completely mixed up in the entire atmosphere.

Use the following basic quantities.

As determined by Avogadro, one atomic weight of any gas contains 6×10^{23} molecules and occupies a volume of 22.4 Liters.

One square inch base area of an atmospheric column weighs 15 pounds.

Air is roughly 80% nitrogen and 20% oxygen. Determine the average atomic weight of air molecules.

9a. Estimate the volume of air a person breathes in liters per second.

9b. Estimate the number of Liters of air Einstein must have breathed in his life.

9c. Estimate the total weight of the earth's atmosphere in grams.

9d. Estimate the total number of molecules in the earth's atmosphere.

9e. What fraction of these molecules are the same as those which Einstein inhaled.

9f. If this is the same fraction of molecules in the air you breathe in, how many molecules of Einstein's air are in your breath?

Exploration Topics

1. Compare the densities of silver, gold, water and air.

2. What happens to the density of water when it turns to ice? Why?

3. Make a simple flow-chart for the evolving views of elements from the Greeks to modern notions.

4. Make some sketches for the different models of the structure of the atom.

5. Look up the relative atomic weights of He, O, N, and Cl (with respect to H). Are these all simple multiples of the atomic weight of H? If not think of a reason for why not?

6. Cite and discuss examples showing the role of cross-cultural exchange in the advancement of knowledge and the progress of science. Cite and discuss examples of the contrary, i.e., restricting the influence of foreign cultures can stifle intellectual growth.

7. Discuss the significance and the impact of the great library at Alexandria. Discuss the parallel role of the world wide web in today's culture.

Sample Quiz

1. Why did Aristotle contend that the laws of nature operating in the heavens must be different from the laws operating on earth?

   a. Because heavenly bodies always appear the same, whereas terrestrial bodies are in constant flux.

   b. Because heavenly bodies move eternally in circles, whereas circular motion is unnatural on earth.

   c. a and b

   d. none of the above

2. What is the same about graphite, diamond and bucky-balls?

   a. They have the same crystal structure.

   b. They are all made of Hydrogen and Oxygen.

   c. They are all made of Carbon atoms.

   d. None of the above.

3. Eratosthenes determined the circumference of the earth to be 40,000 km. What is the radius of the earth (choose the closest answer)?

   a. 13,000 km b. 6400 km

   c. 20,000 km d. 1000 km

4. If the earth circumference is 40,000 km, what is the volume of the earth? Chose the closest answer.

   [Volume of a sphere = $(4/3)\pi R^3$]

   a. 1.08×10^{21} m^3 b. 1.08×10^{25} m^3

   c. 8.65×10^{21} m^3 d. 1.08×10^{12} m^3

5. Convert the earth's density of 5.5 gm/cc to the units of kg/ m^3

   Chose the closest answer.

   a. 1 kg/ m^3 b. 10 kg/ m^3 c. 100 kg/ m^3

   d. 1000 kg/ m^3 e. 10,000 kg/ m^3

6. Suppose the moon's radius is 1/4 the radius of the earth. Think of the moon as a far away ship in space. Estimate the distance to the moon assuming that the moon's radius makes an angle of 0.25 degrees at the eye as shown in the diagram below. (In a right angle triangle with a base angle of 0.25

degrees, the base is 230 times larger than the height.) Choose the closest answer. See figure on Pg 31.

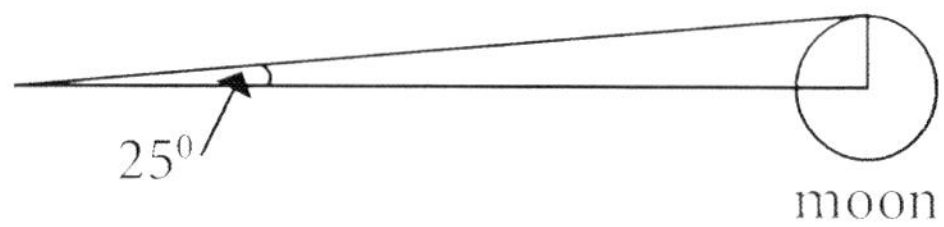

a. 100,000 km b. 200,000 km

c. 300,000 km d. 400,000 km

7. How do we know that water is not an element?

   a. Because water cannot explain the complexity of substances.
   b. Because water can be broken down into the gases hydrogen and oxygen.
   c. Because the Greek ideas of elements are wrong.
   d. none of the above

8. How do we know that light can be treated as a wave?

   a. Because light forms sharp shadows.
   b. Because light can bend around sharp edges such as the edge of razor blades to form a diffraction pattern.
   c. Because white light can be broken down into the colors of the spectrum.
   d. Because each element in a discharge tube gives of discrete color spectrum.

9. What *evidence* led Dalton to propose the notion that the chemical elements are composed of fundamental units he called 'atoms'?

   a. He was influenced by Democritus' ideas of the structure of matter.
   b. He was influenced by Pythagoras ideas of an elegant theory.
   c. He could see the atoms using a scanning tunneling microscope.
   d. He recognized a number pattern in the weight proportions of chemical combinations, such between carbon and oxygen.

10. What evidence led Faraday to propose the idea that all charges are integral multiples of a single fundamental unit of charge, and that one never finds fractions of that charge.

   a. He discovered the electron which is the fundamental unit of charge.
   b. The amount of the chemical element that is evolved during electrochemical decomposition is proportional to a simple fraction x the atomic weight of the chemical element.

c. When two elements A and B combine to form multiple compounds, one atom of A combines with one atom of B to form a binary compound, or one atom of A combines with two atoms of B to form a ternary compound.

d. The atom has both positive and negative charges.

11. How did Rutherford know that the positive charge of an atom resides in a very small volume he called the nucleus?

a. He measured the size of the gold atom.

b. When he bombarded gold atoms with positively charged alpha particles most of them sailed right through, but a very few bounced back.

c. He removed the electrons and measured what was left behind.

d. None of the above.

12. Pick the correct sequence if the entities are arranged in the order of decreasing mass.

a. u-quark, proton, electron, neutrino, hydrogen atom, water molecule

b. proton, u-quark, electron, neutrino, hydrogen atom, water molecule

c. hydrogen atom, water molecule u-quark, proton, electron, neutrino

d. water molecule, hydrogen atom, proton, u-quark, electron, neutrino

e. water molecule, hydrogen atom, u-quark, proton, neutrino, electron

13. Which of the following statements is false?

a. Many chemical elements of the periodic table are found in stars.

b. After the Big Bang, the universe evolved from quarks and leptons into the protons, neutrons and atoms.

c. The search for symmetry, order and unity failed miserably in the quest to understand the order of the periodic table.

d. It is possible to transform one chemical element of the periodic table into another.

e. Mathematical patterns played an important role in comprehending the order and structure at every stage of the elementary quest.

14. Which of the following statements is NOT true about Bohr's model of the atom?

a. Only certain orbits are allowed for an electron around a nucleus.

b. An electron loses energy while it resides in one orbit.

c. When an electron jumps from a higher orbit to a lower orbit it radiates the energy difference of the two orbits.

d. The radii of permitted orbits increases as the squares of the integers.

15. How was Gell-mann's theory of quarks an advance in understanding of elementary particles?

a. Simple combinations of a few quarks could explain the properties of hundreds of baryons and mesons.

b. He linked octet groups of mesons and hadrons to the Buddhist Eight Fold Way

c. He coined the name quarks from James Joyce's *Finnegan's Wake*

d. All of the above.

Answers to Chapter 2 Sample Quiz

1. c
2. c
3. b
4. a
5. e
6. d
7. b
8. b
9. d
10. b
11. b
12. d
13. c
14. b
15. a

Homework Questions for Chapter 2 (with sample answers)

1. What was Aristotle's theoretical explanation that the earth must be in the shape of a sphere? Why does the earth need no support in his theory?
2. Why did Aristotle contend that the laws of nature operating in the heavens must be different from the laws operating on earth?
3. Give two examples (other than the flat earth) of observations that can be deceiving.
4. Explain why Aristotle argued that a heavy body should fall to the earth faster than a lighter body.
5. Why did Aristotle explain comets as atmospheric phenomena, rather than think of comets as heavenly bodies?
6. What asymmetry did Eratosthenes run into in the records of the library? How did he resolve that asymmetry?
7. Redraw Eratosthenes' diagram to compare the shadow made by an obelisk at the equator with the shadows made at Syenne and Alexandria.

Answer

We show in the figure below the obelisks and their shadows in Alexandria, Syenne, and the equator. At noon during the summer solstice the sun is directly above Syenne, so there is no shadow. Because Alexandria is slightly further North, there is a small shadow there which points North. The equator is South of Syenne, so there is a shadow pointing to the South.

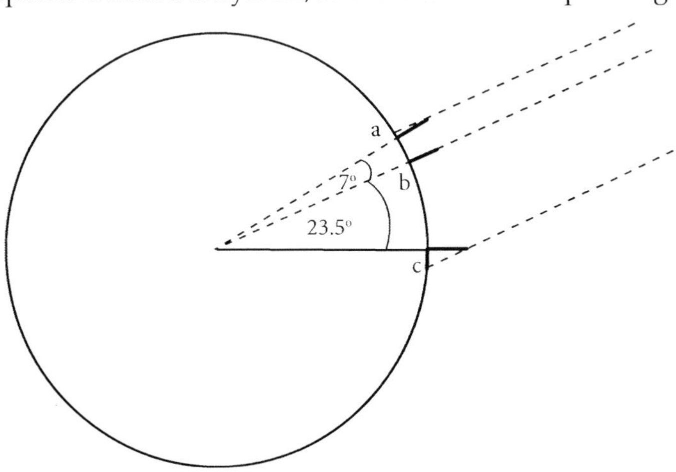

Figure: We represent three obelisks and their shadows as bold lines at Alexandria (a), Syenne (b), and the equator (c). Syenne is about 23.5 degrees North of the equator and Alexandria is about 7 degrees North of Syenne. The rays of the sun are represented by the dashed lines, which are approximately parallel because the Sun is so far away.

8. Are the rays from the sun arriving at two points on earth exactly parallel? If not, why not? Draw a sketch of the earth and sun and rays.

9. Discuss whether it is possible to draw a square with area equal to the area of a given circle.

10. Estimate the volume of the earth in cubic meters. Estimate the surface area of the earth in square meters.

11. Explain how a ship made of iron can float in water even though the density of iron is much larger than the density of water.

12. A column of air one square centimeter in cross-section, measured from sea level to the top of the atmosphere, has a mass of about one kilogram. Estimate the total weight of the earth's atmosphere in kg.

Answer

First we need to estimate the total surface area of the earth (question 10 above), and covert from m^2 to cm^2.

the radius of the earth is $R_E = 6.371\times 10^6\, m$. The surface are then is

$$A_E = 4\pi R_E^{\ 2} = 5.1\times 10^{14}\, m^2$$

$$= 5.1\times10^{18}\ cm^2$$

Mass of atmosphere

$$= 5.1\times 10^{18}\times 1\ kg = 5.1\times 10^{18}\ kg$$

13. If one square inch of the atmosphere weighs 15 pounds, estimate the height of a column of water with base area of one square inch and weighing 15 lbs. The density of water is one gram/cc. That is the maximum height of a water column that the atmospheric pressure will support, so the water column won't fall down.

14. Working with very large numbers. Read about Avogadro's number in the Study Guide text. Estimate the total number of molecules in the earth's atmosphere.

15. How did Dalton estimate the atomic weight of oxygen relative to hydrogen? Why was his result wrong by a factor of more than 2 or more?

16. Giving three specific examples from the quest for the chemical elements, (Lavoisier to Mendeleyev) discuss why close attention to minor details can be important to the progress of science.

17. Is electric charge discrete or continuous? How big is the smallest unit of electrical charge?

18. How did Thomson's discovery change the picture of Dalton's atoms?

19. How did the dogma of indestructibility of atoms arise and how was it destroyed?

20. How do we know that light can be treated as a wave? Compare the wavelength of the different colors of the rainbow. Make a Table of color and wavelength.

Solution

The wave nature of light is explicit from interference experiments. Light waves passed through two spatially separated slits interfere with one another, generating a pattern with varying intensity. Interference experiments result in beautiful patterns, for example http://en.wikipedia.org/wiki/Interference. Visible light has wavelengths ranging from 400nm (violet) to 700 nm (red).

Blue	450 nm
Green	525
Yellow	575
Red	650

21. When radium (atomic number 88) emits an alpha particle, what is the atomic number and the atomic weight of the resulting nucleus.

22. If the atomic weight of He is 4 times the atomic weight of H, why did Rutherford assign only 2 protons to the nucleus of the He atom?

23. Is it possible for the atom of one element to turn into an atom of another element naturally or artificially?

24. Why do all the electrons in an iron atom not settle together into the lowest energy state?

25. How many electrons are found in the outermost shell of Chromium. Explain how you got your answer.

26. How was the Bohr model similar to the model of the solar system? How was it different? Why was it (temporarily) accepted despite its obvious defects? What are the satisfying (dis-satisfying) aspects of Bohr's atomic theory?

Sample Answer

Bohr's model was similar to the model for the solar system because just like the planets revolve around the sun in almost circular orbits, electrons can be thought of as orbiting the nucleus in circular orbits. Instead of being held by the gravitational force, electrons are held together by the electrostatic force between the positively charged nucleus and the negatively charged electrons. The key difference is that electrons can orbit around the nucleus only in orbits of fixed radii, which overcame the paradox that atoms should be unstable since electrons moving in circular orbits lose energy, and hence should collapse into the nucleus. Forcing them to move around in orbits of fixed radii solves the problem. Bohr's model explained the Balmer series of spectral lines for H and predicted other spectral series which were later found. However, the Bohr model did not explain why only discrete orbits were allowed without invoking new ideas from quantum mechanics, such as energy is quantized, or the wave-nature of the electron.

27. Explain the difference between a proton and a neutron in terms of Gell-mann's theory of quarks. Draw sketches.

CHAPTER 3

Reading Questions

P. 139:What was the impact of Rome's ascendancy on Greek culture and science?

How did Greek culture influence Roman culture?

How did Alexandrian culture influence Roman culture?

Give an example of the influence of Greek art on Roman art.

P. 140:In what areas did Romans make advances?

P. 141:Why did the early Christian fathers reject the importance Greek learning?

P. 142:What factors led to the decline of Greek and Alexandrian science?

P. 143:How was Greek science saved from complete obliteration after the fall of the Roman Empire?

P. 144:What was the role of Christianity after the fall of the Roman Empire?

Why were Latin priests uninterested in scientific inquiry?

What was the impact of early Christian attitudes on art?

P. 145:How did astronomy influence Islamic culture?

P. 146:How did Greek culture and science influence Islamic culture and science?

Give an example of the progress of mathematics in Islamic culture.

Give an example of the progress of astronomy in Islamic culture.

P. 148:How did the ideas of symmetry advance in Islamic culture?

What was the role of Christianity during the recovery of European culture after the Dark Ages?

P. 149:List some of the cultural transformations that took place in Europe before the clash with the Islamic culture.

P. 150:How did some of the major works of Aristotle, Plato, Euclid and others finally re-enter Europe after 1000 AD?

Give specific examples how did Islamic advances in mathematics, astronomy and other fields influence European culture?

P. 151:What is the relationship between the Fibonacci series and the Greek Golden Ratio?

How did Islamic scholars like Averroes depart from the teachings of the early Christian theogians?

P. 153:What were some of the troublesome contradictions between the re-discovered Greek ideas and Christian principles?

How did Aquinas avert the emerging crises between reason and faith?

P. 154:What are the parallels between the ideas of Aquinas and Averroes?

How would Aquinas' teachings help with some of the modern controversies: e.g. evolution Vs. creation?

P. 155:What were some of the flaws in the scholastics approach to Greek thought?

P. 156:What elements of the scholastics approach to knowledge inhibited the progress of science?

P. 157:How was Roger Bacon's approach a positive influence on the further development of science?

Why did Roger Bacon's ideas not take hold immediately?

What were the parallels between St Francis' teachings and Roger Bacon's approach?

What positive studies did Bacon conduct in accordance with his fresh approach? What were the immediate benefits of Bacon's empirical studies?

P. 159:What properties of magnets led to the many colorful stories to account for the behaviors?

P. 160:How did Robert Norman characterize the value of an experiment to advance knowledge?

What is magnetic dip?

What is magnetic declination?

How is it possible to impart the property of a magnet to a steel needle which at first does not behave like a magnet?

P. 160:What led Gilbert to propose the idea that the earth behaves like a magnet?

P. 161:How did Gilbert's idea of the earth-magnet explain the orientation, declination and dip of the magnetic needle?

P. 162:How did Gilbert show that the force of attraction of a bar magnet is strongest near the poles?

P. 163:Why is the Ockham's Razor principle attractive for a scientific explanation ?

P. 165:List some of the economic transformations that took place in Europe over the 15th and 16th centuries.

What was the impact of the economic transformations on the power structure?

P. 166:How did the Humanists pose challenges to the established order?

P. 167:How did navigators of the oceans benefit from scientific knowledge?

List some of the technological advances that took place between ca. 1400 to ca. 1500.

What was the impact of the arrival of the printing press?

P. 169:What new approaches developed among the artists of the Renaissance that were later shared by the emerging scientists ?

P. 170:How was anatomist Vesalius' work influenced by new trends developing in the artistic world?

How did Vesalius' work influence artists?

P. 173:How did the importance of the individual return to European culture in the Renaissance?

P. 174:What are some of the important elements of the technique of linear perspective?

What is the vanishing point?

P. 176:Why did daVinci' scientific and technological work fail to have an immediate impact on the scientific development of his time?

P. 177:What were some of the early hints of a link between electricity and magnetism?

P. 178:What was the basic principle underlying the mechanistic philosophy?

How did the Romantic Movement depart from the mechanistic view?

P. 179:Why does the flow of electric current through a wire deflect a magnetic needle nearby?

How does the deflection of a magnetic needle depend on the direction of electrical current flow?

How can you use Oersted's effect to make a detector of electric current (galvanometer)?

P. 181:A permanent bar magnet lies inside a coil of wire connected to a galvanometer. Think of two different ways to produce an electric current to be detected by the galvanometer.

P. 182:How did Faraday make visible the lines of force surrounding a bar magnet?

How is it possible to make visible the lines of force surrounding an electric charge?

P. 186:What is the relationship between Maxwell's electric and magnetic fields and Faraday's lines of electric and magnetic force?

What is the symmetrical relationship between and electric field and a magnetic field?

What were the new predictions from Maxwell's equations unifying electricity and magnetism?

P. 187:What areas of physics did Maxwell's theory unify?

Apart from light, name a few other familiar forms of electromagnetic waves?

Give some examples of imagination and creativity in the development of electromagnetism.

Chapter Questions

1. How did the crusades have a positive impact of the European awakening?
2. Cite one example of the positive impact and one example of the negative impact of Aquinas ideas on the scientific climate.
3. What attitude of St. Francis provided a climate for the growth of science?
4. What aspects of the medieval scholastics approach did Roger Bacon criticize?
5. Why did early medieval artists ignore nature in their paintings?
6. What is the meaning of vanishing point?
7. How did the invention of the printing press contribute to development of science?
8. What are the parallels between the monks of the early middle ages and Egyptian temple-priests in their attitude towards knowledge?
9. What aspects of the Renaissance imply a literal meaning of the word re-birth?
10. Cite two examples of realism that dominates Renaissance art.
11. What are the parallels between Gilbert's and Bacon's approach to advancing science? How was their approach a step beyond the Greek methods?
12. What are the advances of the general empirical method over Aristotle's observation approach? Give examples.
13. Explain the origins of the concept of momentum?
14. How did the development of the technique of linear perspective influence the development of science?

14. Cite a few of Leonardo da Vinci's scientific contributions.

15. How did the Romantic Movement influence the approach of Oersted?
16. What effect led Oersted to establish a link between electricity and magnetism?
17. What effect led Faraday to establish a link between electricity and magnetism?
18. How are Oersted's effect and Faraday's effect related?
19. How did Maxwell take the unification of electricity and magnetism to the next step?

20. How did Maxwell know that light must be a form of electromagnetism?

21. What is the meaning of an electric (magnetic) field?

22. How do you detect the presence of an electric field at a given point in space?

23. How do you detect the presence of a magnetic field at a given point in space?

24. Give some examples of the role of symmetry in the development of electromagnetism.

25. Give some examples of role of unity in the development of electromagnetism.

Exploration Topics

1. Discuss the important factors in the disappearance of the intellectual tradition of Greek rational thought and its eventual re-emergence.

2. In our quest to understand nature, compare the relative merits of knowledge from authority, from reason and from experience.

3. Using examples from early Greek, Alexandrian, Roman, early Christian and medieval Christian cultures, discuss how scientific thought can flourish in the climate of cultural creativity, or can wither in the period of cultural decline.

4. Cite and discuss examples from this chapter showing the role of cross-cultural exchange in the advancement of knowledge and the progress of science. Cite and discuss examples of the contrary, i.e., restricting the influence of foreign cultures can stifle intellectual growth.

5. Discuss the importance of knowledge from authority vs. knowledge from experience as it applies to your own educational experience.

Sample Quiz Questions

1. How did the knowledge of Greek and Alexandrian cultures spread into Europe during the early Middle Ages?

   a. When monks discovered the classical works in Spain translated from Greek into Arabic.

   b. When the Emperor of the Holy Roman Empire decided to release the secret works.

   c. When the Pope decided to allow the monks to travel to Greece.

   d. When the Muslims and European cultures traded freely with each other.

2. What aspect of Muslim cultural attitude had a positive influence on the development of science?

   a. Military power

   b. Faith in the Q'uran

   c. Tolerance to other cultures

   d. All of the above

   e. None of the above.

3. What positive studies did Bacon conduct in accordance with his fresh approach?

   a. He explored the skies with the telescope.

   b. He studied the properties of curved bits of glass.

   c. He carried out dissections of human cadavers.

   d. He made direct measurements of the size of the earth.

4. An example of how art influenced the development of science in the Renaissance is

   a. Painters tried to describe nature in exact detail.

   b. Knowledge of the human anatomy increased through dissection of human cadavers.

   c. Painters used the mathematical technique of perspective in giving their art a three dimensional quality.

   d. All of the above.

5. How is it possible to impart the property of a compass-like magnet to a steel needle which at first does not behave like a magnet?

   a. By touching the needle with the magnet.

   b. By stroking the needle with a magnetic pole in one direction.

   c. By flowing an alternating electric current through the needle.

   d. By hammering the needle.

6. What is the vanishing point?

   a. The point at which a moving body comes to rest.

   b. The highest point reached by a ball thrown upward.

   c. When something disappears from view.

   d. The meeting point of parallel lines on a painting that projects a 3-D scene onto a 2-D canvas.

7. How do you detect the presence of an electric field at a given point in space?

   a. When there is a force on a charge placed at the point.

   b. When a magnetic needle placed at the point deflects into a n-s orientation.

   c. When lightning strikes the point.

   d. When a particle of sawdust placed at the point begins to move.

8. How do you detect the presence of a magnetic field at a given point in space?

   a. When a current flows in a wire placed at the point.

   b. When a magnetic needle placed at the point deflects into a n-s orientation.

   c. When there is a force on a charge placed at the point.

   d. When a steel needle placed at the point becomes magnetized.

Answers to Chapter 3 Sample Quiz

1. a
2. e (closest)
3. d
4. b
5. b
6. c
7. b
8. b
9. c
10. a

Homework Questions for Chapter 3 (with sample answers)

1. What was the impact of Rome's ascendancy on Greek culture and science?

   How did Greek culture influence Roman culture...give a specific example.

   How did Alexandrian culture influence Roman culture...give a specific example.

2. Why did the early Christian fathers reject the importance Greek learning? What factors led to the decline of Greek and Alexandrian science? How was Greek science saved from complete obliteration after the fall of the Roman Empire?

3. How did Greek culture and science influence Islamic culture and science? Give an example of the progress of mathematics in Islamic culture. Give an example of the progress of astronomy in Islamic culture.

4. How did Aquinas avert the emerging crises between reason and faith? What are the parallels between the ideas of Aquinas and Averroes? How would Aquinas' teachings help with some of the modern controversies: e.g. evolution vs creation?

5. How was Roger Bacon's approach a positive influence on the further development of science?

   Why did Roger Bacon's ideas not take hold immediately? What were the parallels between St Francis' teachings and Roger Bacon's approach?

6. Why did early medieval artists ignore nature in their paintings?

7. What is the meaning of vanishing point? How did artists use the vanishing point? Draw a sketch.

   Sample Answer

   In order to trick your mind into thinking a two-dimensional image is a three dimensional scene, one important trick that mathematicians and artists discovered is the vanishing point. Things in the distance appear to get smaller and closer together, such as two parallel train tracks viewed over a long distance. To represent this effect in a two-dimensional image, artists artificially 'squeeze' objects that are meant to be further away by having any parallel lines in their subject meet at a point on the painting (in real life parallel lines never cross). This point appears to your eye to be very far in the distance. Thus parallel lines in 3D converge to a single "vanishing point" in a 2D representation.

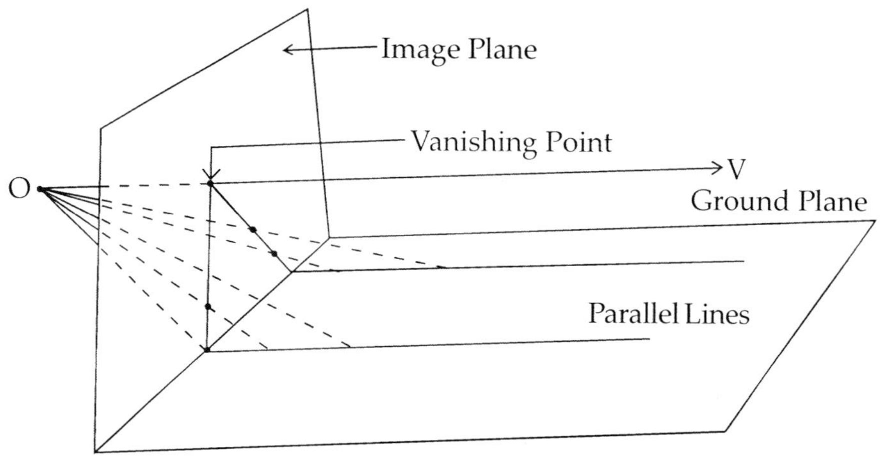

(http://en.wikipedia.org/wiki/Vanishing_point).

8. Describe the many phases of the changing relationship between Faraday and Davy.

Sample Answer

Faraday started as a bookbinder and attended Davy's lectures. He made a book out of his notes and presented it to Davy asking him for a job. Davy gave him a job to wash bottles. Later Davy made him his assistant for chemical experiments. But Davy never treated Faraday with any respect. It is often said that Davy's greatest discovery through his career was Faraday!

9. What were some of the early hints of a link between electricity and magnetism? Why does the flow of electric current through a wire deflect a magnetic needle nearby? How does the deflection of a magnetic needle depend on the direction of electrical current flow?

10. A permanent bar magnet lies inside a coil of wire connected to a galvanometer. Think of two different ways to produce an electric current to be detected by the galvanometer.

11. How did Faraday make visible the lines of force surrounding a bar magnet? How is it possible to make visible the lines of force surrounding an electric charge? How does an electric field influence a magnetic field a vice versa?

CHAPTER 4

Reading Questions

P. 190: Why were the faculty at the University of Pisa not convinced by Galileo's experiment showing Aristotle's mistake?

How did the re-discovery of classical knowledge influence Galileo's work?

P. 191: How did the approach of the Renaissance artists influence Galileo's approach to science?

P. 192: What practical application emerged from Galileo's pendulum discovery?

What were some of the surprising aspects of Galileo's discoveries about pendulum swings?

P. 195: How was Galileo influenced by Parmenides' and Plato's approach to understanding nature?

How was Galileo influenced by Aristotle's approach to understanding nature?

P. 196: List three practical applications of air resistance.

What was the influence of the Humanists on Galileo's approach and methods?

P. 197: How did Galileo depart from the Renaissance world view?

What are the parallels between Galileo's approach to science and the Baroque artist approach?

P. 198: How did the Baroque artists transcend the approach of the Renaissance?

P. 199: How was Galileo's work on dynamics influenced by Pythagoras, Aristotle, Socrates, Plato, and Bacon?

P. 200:What new concepts did Galileo introduce in his analysis of motion?

P. 201:What mathematical pattern did Galileo find in his observation of free fall motion down an incline plane?

P. 202:What are the definitions of "velocity" and "acceleration"?

For constant acceleration, what are the relationships between velocity and time, between total distance traveled and time?

P. 204:Describe the motion of a ball launched upward with an initial velocity in terms of its acceleration, velocity, and distance traveled before it reaches the top of its excursion?

What are the ideal laws of free fall motion?

Why are these laws called "ideal"?

Describe the effect of air resistance on the velocity of a free falling body?

P. 205:What are the physical factors which influence terminal velocity of a falling body?

P. 206:How did Galileo use extrapolation to arrive at underlying ideal laws?

Why does an object moving along a horizontal plane lose speed?

What is the ideal motion for a body moving along a horizontal plane?

Describe the ideal motion of a body in free fall after it hits the ground?

How did Galileo's treatment of free-fall motion unify all types of vertical motion?

P. 207:Give a formal definition of inertia.

Give another example from horizontal motion of the consequences of inertia.

Give another example from the development of science for Einstein's maxim: 'Imagination is more important than knowledge.'

P. 208:How did the approach of the Renaissance artists influence Galileo's approach to discovering the parabolic nature of the projectile's trajectory?

P. 210:What is the magnitude of the vertical component of the velocity of a projectile at the top of its parabolic path?

How does the magnitude of the horizontal component of the velocity of a projectile change over the course of its ideal parabolic motion?

What should be the angle of the cannon tube with respect to the horizontal for the projectile to stay in the air for the longest time?

P. 212:What was the key difference between the gunners' approach and Galileo's approach to the knowledge about the angle to set the cannon for maximum range?

What were the consequences of Galileo's approach?

P. 213:What are Newton's definitions of "force" and "mass"?

P. 214:What is the relationship between "mass" and "inertia"?

What is the difference between "mass" and "weight"?

P. 215:What is the distinction between "kinetic energy" and "potential energy"?

In what part of the swing does the pendulum bob have the maximum kinetic energy?

In what part of the swing does the pendulum bob have the maximum potential energy?

P. 216:What is the effect of time reversal symmetry on "velocity" and on "acceleration"?

P. 217:Suppose you are watching a video of a ball bouncing off the floor in vertical motion? If the motion is ideal (no loss of energy due to friction) and the bounce perfectly elastic (no loss of energy) can you tell if the video is playing forward or backward?

P. 218:What is the effect of mirror symmetry on three velocity components of motion of a ball positioned in front of a mirror?

How should you orient a clock in front of a mirror so that the clockwise motion of its hands appears as counterclockwise motion?

What is the difference between a "mirror transformation" and a "parity transformation"?

P. 220:In the example of the observer on the ground observing the motion of the ball launched on a moving ship, how would the person on the ship have to launch the ball so that the observer on the ground would see the ball moving in a straight up-down trajectory?

P. 221:Imagine you are on an airplane cruising at a constant speed of 500 mph. Compare the experience of pouring wine from a bottle to a glass with the same experience when the plane was parked at the gate?

P. 222:Why were "lecture demonstration experiments" considered unnecessary by some of the other professors at Padua?

P. 223:What are the different aspects of observation that are important for the progress of science?

What is the general significance of an asymmetry found in a physical observation?

Think of some examples.

What is broken symmetry? Think of some examples.

What general symmetries are always valid in the laws of nature?

What special symmetries are sometimes found violated?

P. 224:What is the difference between symmetry breaking due to outside interactions versus spontaneous symmetry breaking? Think of some examples.

P. 225:Which forces of nature violate parity conservation?

Which force is responsible for nuclear decay?

P. 226:Give two examples of parity violation in the nuclear/sub-nuclear world.

Give an example of charge-conjugation (C) violation.

P. 227:Give two examples of CP symmetry in weak interactions.

P. 228:What is the nature of the matter-antimatter asymmetry in the universe?

What is the impact of the matter-antimatter asymmetry in the universe?

P. 229:How are electricity and magnetism related to each other to show an underlying unity? (review updates of chapter 3).

How did Maxwell arrive at the conclusion that light must be a wave of changing electric and magnetic fields? (review updates of chapter 3).

P. 230:Why did Einstein postulate that the speed of light in vacuum must be a constant of nature?

How does the constancy of the speed of light lead to a paradox?

What were some of the early influences that led Einstein to think "out of the box"?

P. 231:What would be the consequences to familiar devices based on electromagnetics if the speed of light would change with the speed of a moving vessel?

P. 232:The speed of sound in air is 340 m/s. What would be the speed of the sound wave measured by a navigator on a ship moving toward a fog horn at 10 m/s?

P. 233: Einstein realized that in order to obtain the same number for the speed of light in all moving vehicles (independent of their speeds) certain fundamental aspects of the way we think about reality are seriously flawed and must be corrected. What aspects of reality was he referring to?

P. 234:If clocks slow down on a fast moving space-ship will it take longer to go through for a 2 hour movie on the space-ship than at home on earth?

P. 235:If the length of your space capsule measures 100 m with a meter ruler on board while stationary on earth, will the length you measure with the same meter ruler be shorter, or still 100 m, when the space ship is moving toward Mars at high speed?

P. 236:If two space ships are moving toward each other at speeds 0.5 c each relative to a stationary observer, what will be the speed of one ship measured by the other approaching ship?

What will be the distance between the sun and the star Sirius from an imaginary space-ship moving at speed c toward Sirius?

P. 237:Is it possible for two observers on spaceships moving at different speeds to disagree about the time-order of two events that occur in the same place?

P. 238:Why is the twin paradox between the stay-at-home twin and the space-traveler twin not symmetric between the twins?

P. 239:How many meters does a light beam travel in one nanosecond (billionth of a second) ?

P. 240:Is it possible to convert energy into matter? If yes, give an example of how this can be done.

P. 241:Besides upending the everyday, conventional views of absolute space and absolute time, what other conventional notions did the special theory of relativity uproot?

Math Based Topics

Velocity as a Function of Time

We now redo Galileo's definitions and derivations using modern terms and algebra to develop some general results. We will see how formal mathematics enables us to imagine more than we can clearly think with just strings of words. Thinking about physical quantities in terms of numbers helps to take some of the mystery out. We will also see how the symmetries of physical behavior reflect in the mathematical equations and graphs of the physical laws that describe motion.

The physical quantity, *velocity*, is defined as the ratio of two other physical quantities: *distance traveled* and *time elapsed*.

$$\text{Average velocity (v)} = \frac{\text{distance interval (d)}}{\text{time elapsed (t)}}$$

Distance interval	1	3	5 inches
Total distance (D) at time t	1	3	9
t_{mid} =	0.5	1.5	2.5 sec
Av. Velocity (v) at time t_{mid}	1	3	5 in/sec

Fig. 4.1 When an object moves with uniform acceleration down an inclined plane, the velocity increases in proportion to the elapsed time. In the upper box, we mark the time elapsed and list the distance intervals after each interval of elapsed time. In the lower box we calculate the velocity for each interval. Since time is always flowing we need to assign a time value to every time interval. We could use the start time of the interval, or the end time, or the average time. We decide to use the average time to label the first, second, third... intervals: t = 0.5, 1.5, 2.5...

$$\text{acceleration} = \frac{\text{change in velocity}}{\text{time interval}}$$

t	0.5	1.5	2.5 sec
V=2t	1	3	5 (in/sec)
Change in velocity	2	2	2(inches/sec)

Acceleration=2 inches/sec/sec=2 in/sec^2, Acceleration is constant

Fig. 4.2: For uniform acceleration, the change in velocity is the same for equal intervals of time. To calculate acceleration, we examine the change in velocity for each interval and divide by the time interval.

Units

It is very important to be careful about the units we decide to use, and to use them consistently throughout the analysis. If we change the units we must be careful to apply the right conversion factors. Using the wrong units can be fatal! An airline pilot who receives orders from the control tower in feet should not set his altitude based on meters, or he may collide with another plane circling the airport! When refueling an airplane it is crucial to fill the right amount of *gallons* or *liters* as called for in the instruction manual of the aircraft.

There are many units of distance, such as: inches, feet, kilometers, miles, ...

There are also many units of time , such as: seconds, minutes, hours,...

The units for length (meters) and time (seconds) in scientific work are from the Scientific International (SI) system. Wherever possible it is advisable to use this system.

Velocity Examples

A car is traveling at a uniform velocity of 5 m/sec toward north. How far from the starting point will the car travel in 20 seconds?

$$\text{velocity} = \frac{\text{change of distance}}{\text{change of time}} \quad \text{Change of distance} \Rightarrow \text{velocity} \times \text{time elapsed}$$

$$\text{change of distance} = \text{final distance - initial distance} = \text{distance traveled}$$

$$\text{distance traveled} = 5\left(\frac{m}{\text{sec}}\right) \times 20\ (\text{sec}) = 5 \times 20 \quad \frac{m}{\text{sec}} \times \text{sec} = 100\text{m}$$

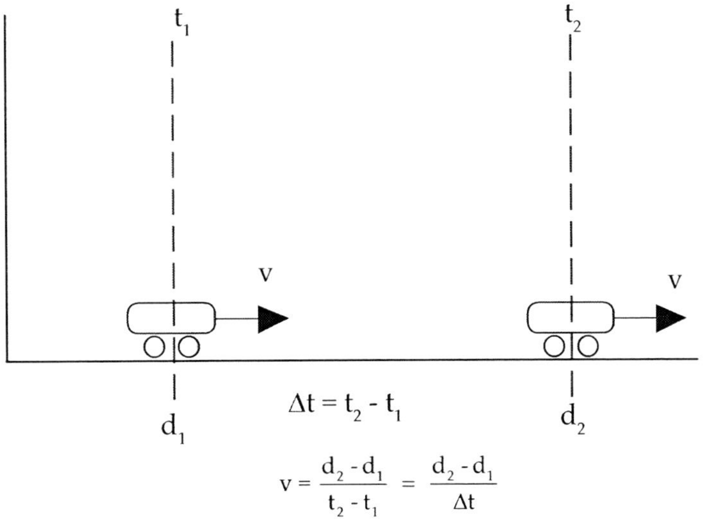

Movie running backward

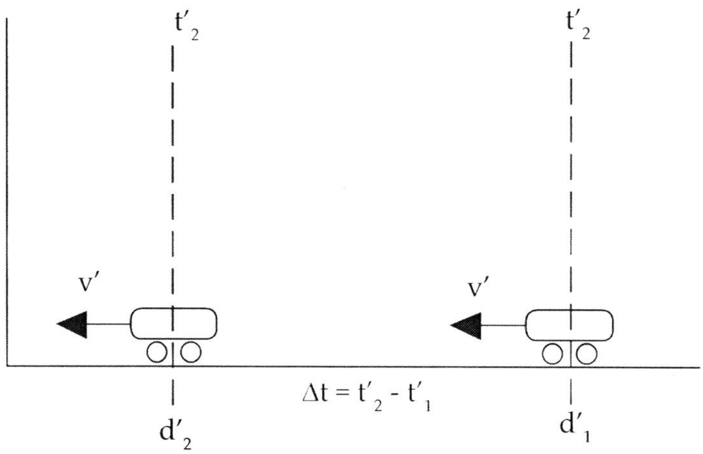

$$v' = \frac{d'_2 - d'_1}{t'_2 - t'_1} = \frac{d_1 - d_2}{\Delta t} = -v$$

Fig. 4.3: Effect of time reversal on velocity using the definition of velocity.

Effect of Time Reversal on Velocity

Suppose a car is moving at a velocity v. We make a video recording of the car's motion and play it backward in time. On screen, it would appear as if the car is going in reverse with velocity (- v). Fig. 4.3 shows how with the formal definition of velocity we reach the same conclusion.

What is even more interesting is that we can reach this conclusion *in a much simpler way* - by just reversing the direction of time flow and substituting (-t)for (t) in the definition of velocity.

$$v = \frac{d_2 - d_1}{t} \qquad v' = \frac{d_2 - d_1}{t} = -v$$

Velocity changes sign under time reversal. Of course, time never flows backward, but thinking about time flowing backward is a very helpful idealization to understand the fundamental laws of motion.

We repeat the analysis using graphs. Figure 4.4 shows what happens to the *distance vs. time graph* for the two cases of motion in uniform velocity, and for the movie playing in reverse. Note that velocity turns out to be the slope of the straight-line graph. Again, we can obtain the answer for the case of the movie playing in reverse by using time reflection symmetry.

$$v = \frac{d_2 - d_1}{t_2 - t_1} = \text{splope of line} \qquad v' = \frac{d'_2 - d'_1}{t'_2 - t'_1} = -v$$

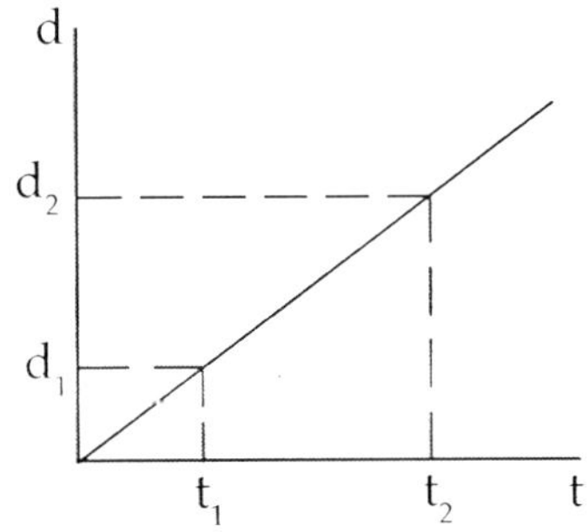

Distance vs. time graph for object moving with uniform velocity. The slope, i.e. the velocity is positive.

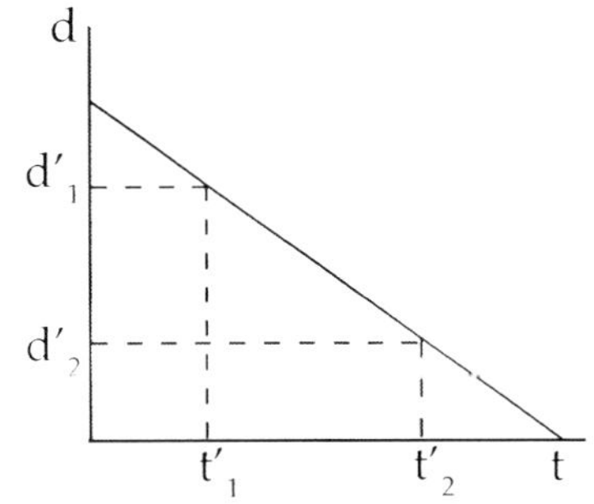

Distance vs. time graph for movie playing backward. Now the slope, i.e. the velocity is Negative.

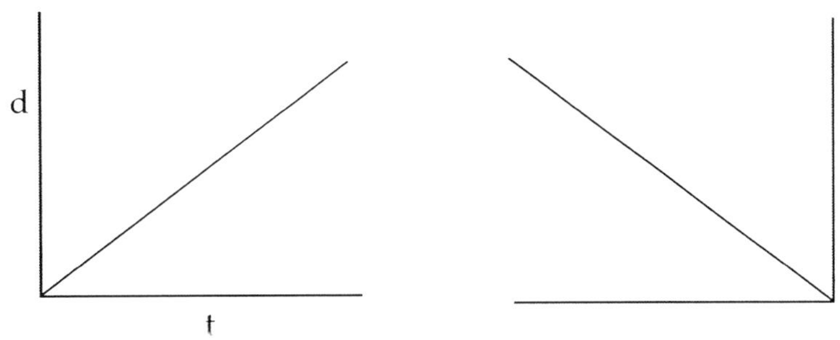

We can obatain the answer of d vs t for the movie playing backward using symmetry, if we reverse the flow of time by making a mirror image of the d vs t graph.

Fig.4.4: Using symmetry to determine the effect of time reversal on velocity.

Acceleration Examples

A car starts from rest and accelerates at 10 m/sec^2. What will be the velocity of the car after 20 seconds?

$$\text{acceleration} = \frac{\text{change of velocity}}{\text{change of time}}$$

$$\text{Change of velocity} = \text{acceleration} \times \text{change of time}$$

$$\text{Change of velocity} = \text{final velocity - initial velocity}$$

$$\text{Now, initial velocity} = 0 \text{ because the car starts from rest}$$

Therefore, change of velocity = final velocity

Change of velocity = final velocity = acceleration x time elapsed

$$\text{Change of velocity} = 10 \ \frac{\text{m}}{\text{sec}^2} \times 20 \text{ sec} = 10 \times 20 \quad \frac{\text{m}}{\text{sec}^2} \ \text{sec} = 200 \ \frac{\text{m}}{\text{sec}^2}$$

Effect of Time Reversal on Acceleration

Suppose a car moving at a velocity v_1 increases its velocity with uniform acceleration *a* to reach a final velocity v_2. Once again, we video record the car's motion and play it backward. It would appear as if the car is going in reverse with large velocity (v_2), then slows down to a final velocity $(-v_1)$. Fig. 4.5 shows how we determine the acceleration in the reverse-play case with the formal definition of acceleration. We must be careful with sign manipulations.

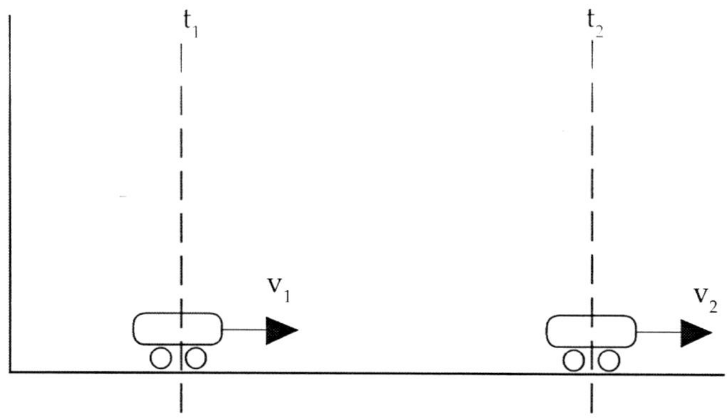

$$a = \frac{v_2 - v_1}{t_2 - t_1} = \frac{v_2 - v_1}{\Delta t}$$

Movie running backward

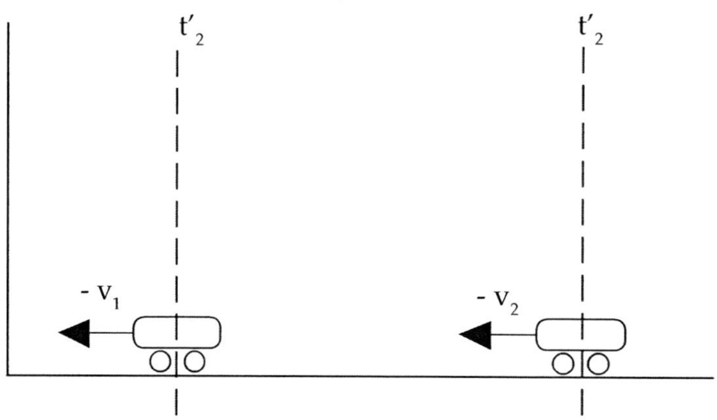

$$a' = \frac{v_f - v_i}{\Delta t} = \frac{-v_2 - (v_1)}{\Delta t} = \frac{v_2 - v_1}{\Delta t} = a$$

Fig. 4.5: Effect of time reversal on acceleration using the definition.

The result is that the *acceleration is the same* for the movie running in reverse. Again, we can reach the same conclusion in a much simpler way, directly from the equations if we just change the sign of the time and substitute (-t) for (t) and (- v) for (v) by ideally reversing the flow of time.

$$a = \frac{v_2 - v_1}{t_2 - t_1}$$

$$a' = \frac{-v_2 + v_1}{-t_2 + t_1} = \frac{v_1 - v_2}{t_1 - t_2} = \frac{v_2 - v_1}{t_2 - t_1} = a$$

Time reversal flips the sign of velocity, but not the sign of the acceleration. Another way to think about this is: Acceleration is the change of velocity, which means the *change of the change* of position. The two sequential changes leave the sign of the acceleration unchanged. It is also important to note that distance and time are independent, so that time reversal does not effect the sign of the distance traveled.

We can revisit the analysis once again using graphs. Fig. 4.6 shows the graph of v *vs.* t for the actual motion and the answer for the movie played in reverse.

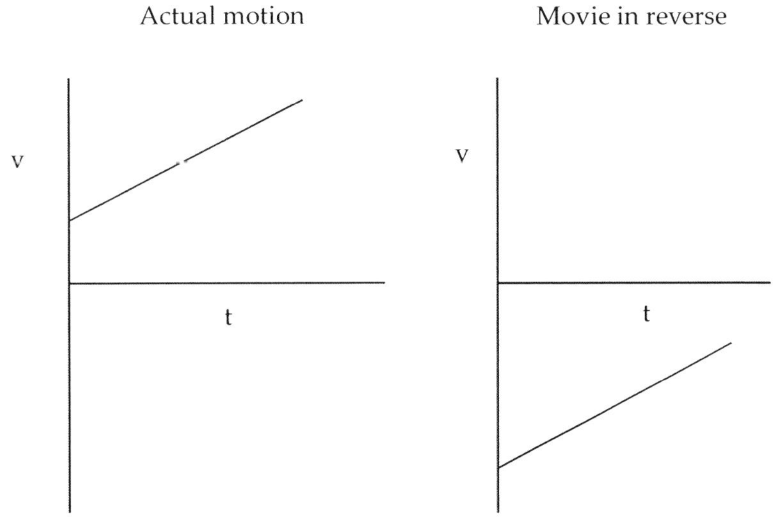

Fig. 4.6: A graph of the effect on time reversal on acceleration.

Another way to get the answer for acceleration in the case of the movie in reverse is by using symmetry as in Fig. 4.7. We first apply time reversal by reflection about the time-axis. In a second step, we apply velocity reversal by reflecting about the velocity-axis. Note that the sign of the acceleration, which is the slope of the velocity vs time graph, remains unchanged after the two sequential reflections.

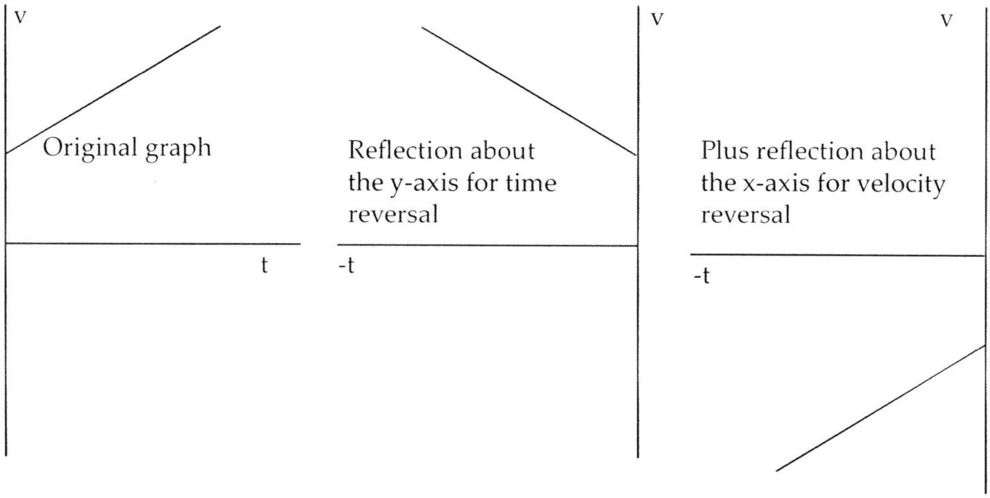

Fig. 4.7: Using symmetry to determine the effect of time reversal on acceleration.

General Equations of Motion for Uniform Acceleration

Our final objective is to derive general equations for the case of uniform acceleration. We plan to derive three equations

1. Velocity vs. time.
2. Distance traveled vs time.
3. Velocity vs. distance.

Velocity vs. Time

Suppose the motion starts at initial time t_i with initial velocity, v_i. We aim to calculate the final velocity, v_f

Using the formal definition of acceleration:

$$a = \frac{\text{change in velocity}}{\text{change in time}}$$

$$a = \frac{v_f - v_i}{t_f - t_i}$$

v_f means final velocity v_i means initial velocity

t_f means final time t_i means initial time

If initial time, $t_i = 0$, as it is at the start of a stopwatch, the acceleration simplifies to

$$a = \frac{v_f - v_i}{t_f}$$

To further simplify the notation, we call $t_f = t$, for the general time

$$a = \frac{v_f - v_i}{t}$$

We have now arrived at the first general equation, or the first equation of motion, for an object undergoing uniform acceleration

$$v_f = v_i + at$$

Recall that uniform acceleration means a constant value of *a*.

An important special case of the general equation occurs when the initial velocity = 0, i.e., $v_i = 0$, as when an object falls, starting from rest. In this case so that the original relationship simplifies to

$$v_f = at$$

Fig. 4.8 gives graphical representations of (a) the uniform acceleration (b) velocity as a function of time for arbitrary initial velocity and (c) velocity vs. time when initial velocity equals zero.

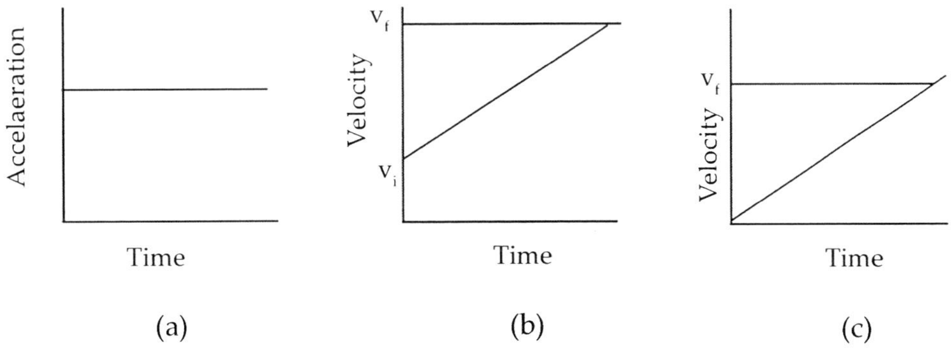

Fig. 4.8: (a) Acceleration is constant. (b) Velocity increases linearly with time. The initial velocity v_i is the y-intercept. (c) In a special case, the initial velocity is zero.

Distance as a Function of Time

Next we derive general equations for the distance traveled after time t. This involves a higher level of mathematics, i.e., calculus. To get around it, we approach the problem in steps. Before we consider uniformly accelerated motion, we approach some simpler cases. We calculate the distance traveled for:

1. Motion with uniform velocity
2. Motion with a step change in velocity
3. Motion with uniform change in velocity, i.e., uniform acceleration, starting with a finite initial velocity.

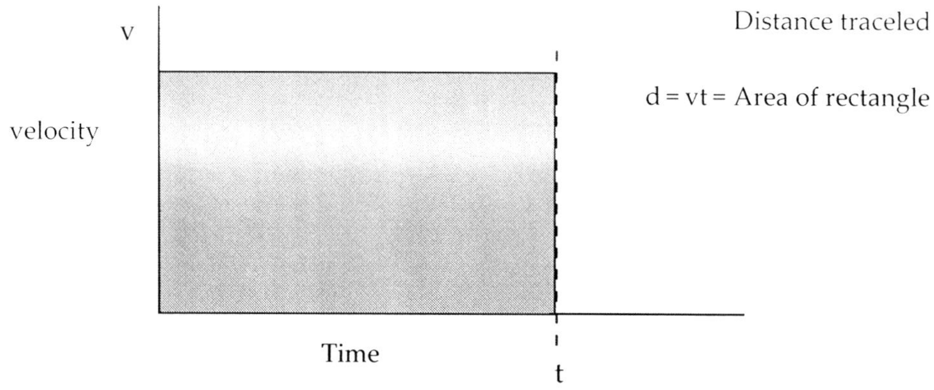

Fig. 4.9: Case 1

Case 1 If a body moves with constant velocity, its velocity-time graph is given. The distance traveled in time t is given by the simple formula $d = vt$, which is the same as the area of the rectangle, or the area below the velocity-time graph between the start and the end times.

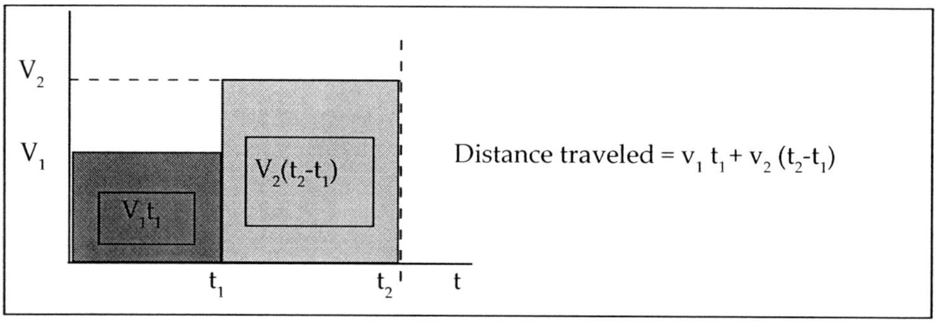

Fig. 4.9: Case 2.

Case 2 If a body moves with constant velocity v_1 from time 0 to time t_1, and then, suddenly, its velocity changes stepwise to v_2. The body moves with velocity v_2 between times t_1 and t_2. Now the distance traveled for each velocity is the area of the individual rectangles. Therefore the total distance traveled is the sum of the areas.

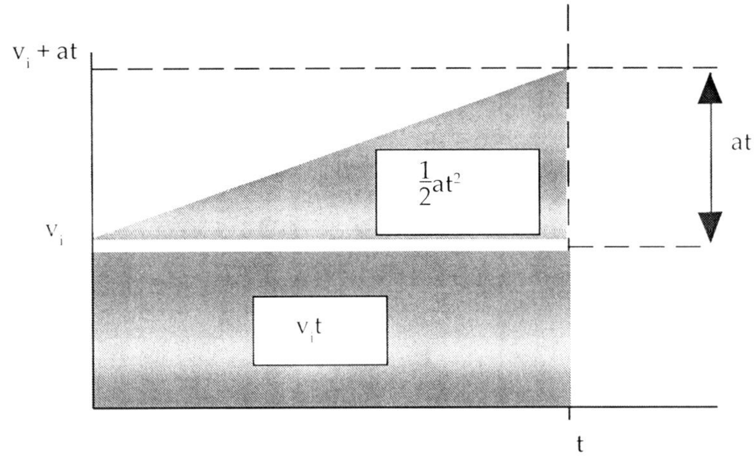

$$d = v_it + \frac{1}{2}at^2$$

$$\text{If } v_i = 0 \qquad d = \frac{1}{2}at^2$$

Fig. 4.9: Case 3.

Case 3 (Fig. 4.9). Finally, consider a body moving with uniform acceleration, which means its velocity is increasing uniformly, or linearly, from an initial velocity v_1 to a final velocity v_2. Again the total distance traveled is the area under the velocity-time curve, which is the same as the area of the rectangle plus the area of the triangle.

$$d = v_it + \frac{1}{2}at^2$$

Example

A car is moving at a velocity of 10 m/sec when the driver decides to overtake another car and accelerates at 5 m/sec² for 10 seconds. After the 10 seconds, he decides to resume uniform velocity. What is the distance traveled by the car after it returns to constant velocity? What is the final velocity?

$$v_i = 10\frac{m}{sec}$$

$$a = 5\frac{m}{sec^2}$$

$$v = v_it + \frac{1}{2}at^2$$

Velocity As a Function of Distance

Finally, we wish to calculate the velocity of a uniformly accelerating body as a function *only of distance*. We want to get rid of time in the equations, because we wish to express how velocity of an object changes with distance. This involves some algebraical footwork.

By the definition of acceleration

$$a = \frac{v_f - v_i}{t} \quad \text{to} \qquad t = \frac{v_f - v_i}{a}$$

Eliminate t from the distance-time equation:

$$d = v_i t + \frac{1}{2} a t^2$$
$$d = v_i \frac{(v_f - v_i)}{a} + \frac{1}{2} a \frac{(v_f - v_i)^2}{a}$$

Now from high school algebra we know that:

$$(v_f - v_i)^2 = v_f^2 - 2 v_f v_i + v_i^2$$
$$d = \frac{v_i v_f - v_f^2}{a} + \frac{1}{2} a \frac{(v_f^2 - 2 v_f v_i + v_i^2)}{a^2}$$

Collecting together similar terms

$$d = \frac{1}{a}[v_i v_f - v_i^2 + \frac{1}{2} v_f^2 - v_f v_i + \frac{1}{2} v_i^2]$$
$$d = \frac{1}{a}[\frac{1}{2} v_f^2 - \frac{1}{2} v_i^2]$$
$$2ad = v_f^2 - v_i^2$$
$$v_f^2 - v_i^2 = 2ad$$

Example

A ball is released from rest from the top of a tower 100 meter high, and accelerates in free fall at 9.8 m/sec². What is the velocity of the ball when it hits the ground?

$$d = 100\text{m}$$
$$a = 9.8 \frac{\text{m}}{\text{sec}^2}$$
$$v_i = 0$$
$$v_f = ?$$
$$\text{Use: } v_f^2 - v_i^2 = 2ad$$

$$v_f^2 = 0 + 2 \times 9.8 \, \frac{m}{\sec^2} \times 100m$$

$$v_f^2 = 1960 \frac{m^2}{\sec^2}$$

$$v_f = 44.3 \frac{m}{\sec}$$

Summary of Laws of Motion Under Constant Acceleration

When a body moves with uniform acceleration (a) as in a free fall or down an inclined plane, the equations of motion are:

$$v_f = v_i = at$$

$$d = v_i t + \frac{1}{2} at^2$$

$$v_f^2 - v_i^2 = 2ad$$

There are important symmetries underlying these equations.

1. The equations can be used to describe uniformly accelerated motion anywhere. It does not matter whether the motion takes place in Pisa or Padua. Just as the pattern of dancing soldiers on Plate 1.6 does not change upon translation, the equations are invariant under translation in space. The formal proof of this symmetry involves higher mathematics than our present scope allows, but we can offer a mathematical proof of another very important symmetry below.

2. The equations of motion are symmetric with respect to time reversal. The same equations can be used to describe vertical motion under uniform acceleration for a movie playing in reverse. We can easily check this by substituting (-t) for (t) and changing the signs of the velocities, but not the sign of the acceleration.

$$v_f = v_i + at \\ \equiv \\ -v_f = -v_i + a(-t)$$

$$d = v_i t + \frac{1}{2} at^2 \\ \equiv \\ d = -v_i (-t) + \frac{1}{2} a(-t)^2$$

$$v_f^2 - v_i^2 = 2ad \\ \equiv \\ (-v_f)^2 - (-v_i)^2 = 2ad$$

Physicists believe that all the fundamental laws of physics obey time reversal symmetry, in the same way as the equations of motion for uniform acceleration obey time reversal symmetry. But we must not take the general principles of symmetry for granted. They must always be subject to experimental test and verification. Any deviation from perfect symmetry uncovered generally implies that there is something incomplete in our understanding. For example, we will soon see how the ideal symmetry in the above equations fails, because the equations do not include the physics of air resistance.

Dimensional Analysis

This is an extremely powerful physicist's tool, one which we will use frequently to keep check on the validity of a mathematical conclusion. In the mathematical expression for distance traveled by an object moving with uniform velocity

$$s = vt$$

The units on both sides of the equation are consistent

$$s = vt$$

$$L = \frac{L}{T} \times T = L$$

We denote the basic dimensions as: length (L) and time (T). On the left hand side, the units of distance are L for length. On the right hand side, the units of velocity are distance/time, i.e., L/T, often written as : LT^-1).

Example

A stone is thrown straight up. After 20 sec have elapsed, it drops back to the ground.

What height does the stone reach?

What is its initial upward speed?

What is its speed upon returning to the ground?

Acceleration due to gravity $g = 9.8 \text{ m/sec}^2$

The solution to this problem can be greatly simplified if we employ the principle of symmetry which we have shown: *The laws of mechanics do not depend on the direction in which time flows.* Of course we never observe time to flow backwards, but since we are considering the ideal laws of mechanics, we may *imagine* time to flow backwards. As before, one way to think about the flow of time is to imagine a movie recorded during the falling part of the stone's motion.

If the movie is then played in reverse, the stone will appear to be projected upward. The initial velocity will be equal in magnitude (but opposite in sign) to the final velocity at which the stone hits the ground in the original version of the movie. On the replay (time running in reverse), the stone decelerates under gravity. It comes to rest at the top of its motion at the end of the same time (t) that it took to hit the ground in the original version. Later, we show with a detailed algebraic proof that the time taken for the two parts of the motion are indeed identical. We will see that the price of ignoring symmetry is more intricate mathematics.

Divide motion into two parts

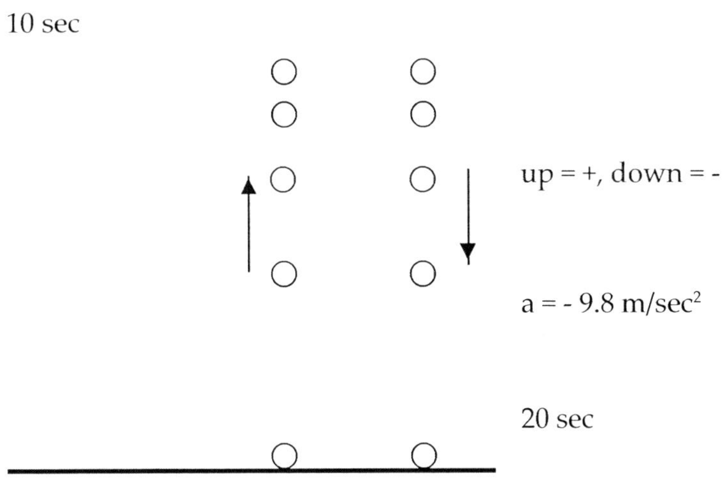

In the first part,	In the second part,
$v_f = 0$, $t = 10$ sec, cals. v	$v_i = 0$, $t = 10$ sec, cals. v_f
$v_f = v_f = v_i + at$	$v_f = v_i + at$
$0 = v_i - (9.8)(10)$	$v_f = 0 - (9.8)(10)$
$0 = v_i - 9.8$	
$v_i = 98 \text{m/sec}$	$v_f = - 98 \text{ m/sec}$

Calculate the distance traveled up

$$2ad = v_f^2 - v_i^2$$

$$2(-9.8)\,d = 0 - 98^2 \qquad d = \frac{98 \times 98}{2 \times 9.8} = 490 \text{ m}$$

Now we prove, in the general case - without specific numbers - that the time taken to rise to the top equals the time taken to descend to the ground. This feature is inherent in time reversal symmetry of the equations of motion

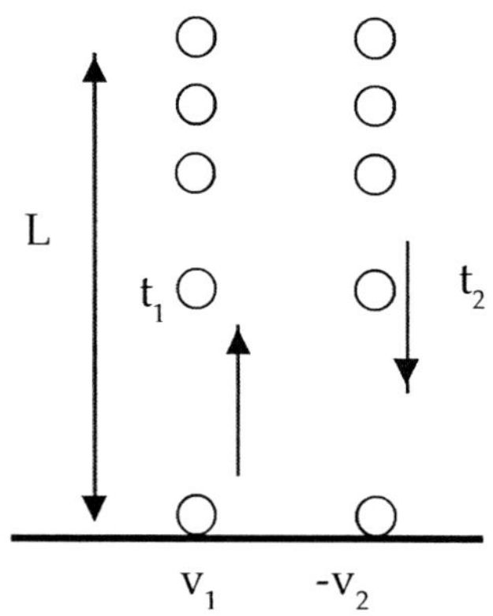

up = +, down = -

$a = -g \text{ m/sec}^2$

downward acceleration

In the first part,

$v_f = 0$, $v_i = v$, calculate t_1

$v_f = v_i + at$ Equation of Motion

$0 = v_1 - gt_1$

$t_1 = \frac{v_1}{g}$

In the second part,

$v_i = 0$, $v_f = v$, 2 calculate t_2

$v_f = v_i + at$

$-v_2 = -g\,t_2$

$t_2 = \frac{v_2}{g}$

Calculate the distance traveled up (L)

$v_f^2 - v_i^2 = 2ad$ Equation of Motion $v_f^2 - v_i^2 = 2ad$

$-v_1^2 = 2(-g)L$

$v_1^2 = 2(g)L \qquad\qquad v_2^2 = 2gL$

$$L = \frac{v_1^2}{2g} \qquad \frac{v_1^2}{2g} = \frac{v_2^2}{2g} \qquad L = \frac{v_2^2}{2g}$$

$$v_1 = \pm v_2$$

$v_1 = -v_2$ Choice made from physical situation

$$t_1 = \left|\frac{v_1}{g}\right| = \left|\frac{v_2}{g}\right| = t_2 \qquad t_1 = t_2$$

Motion in Two Dimenssions

The mathematical equation that describes Archimedes' parabola is

$$y = a\ x^2$$

Suppose the uniform horizontal velocity = v (*a constant*)

Then the horizontal distance traveled is: $x = vt \Rightarrow t = \frac{x}{v}$

Suppose that the body starts out with initial vertical velocity = 0

which means that vertically it is starting from rest.

The vertical distance due to free fall is g *(a constant)*

The vertical distance traveled = $y = \frac{1}{2}\ g\ t^2$

Eliminate t

$$y = \frac{1}{2}\ g\ \frac{x^2}{v^2}$$

$$y = \left(\frac{1}{2}\ \frac{g}{v^2}\right) x^2$$

Which is the equation of a parabola!

The quantity in parenthesis is a constant because both g and v are constants.

Example

A stone is thrown horizontally straight out of an open window of a house. If its initial speed is 20 m/sec, and it hits the ground 2 sec later,

a. How far away from the building will it hit the ground?

b. What is the height at which the stone was launched?

c. What is the final vertical speed of the stone?

d. What is the final horizontal speed of the stone?

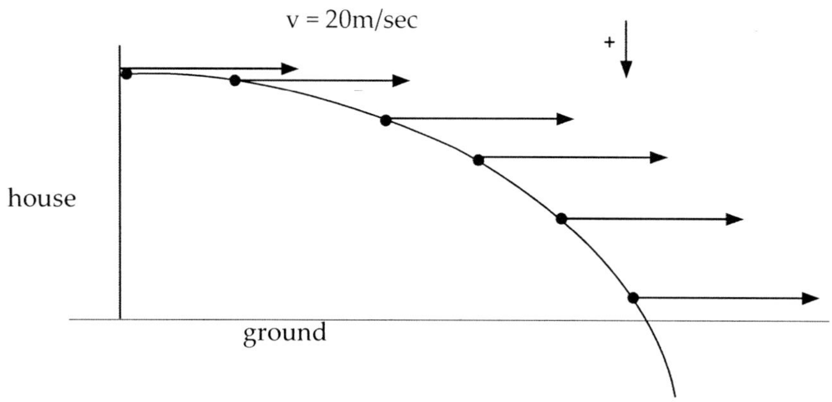

Fig. 4.10

a. <u>Horizontal motion</u>

$t = 2$ sec, $v_h = 20$ m/sec

v_h = horizontal distance/time

horizontal distance = v_h x time

horizontal distance = 20 x 2 = 40 meters

b. <u>Vertical motion</u>

Initial velocity = 0, $v_i = 0$

$d = v_i \times t + (1/2)\, a\, t^2$

$a = g = 9.8$ m/sec^{2,} $t = 2$ sec

$d = 0 \times t + 1/2 \times 9.8 \times 2^2$

$d = 2 \times 9.8 = 19.6$ m

c. $v_f^2 - v_i^2 = 2\, ad$

$v_f^2 - 0 = 2 \times 9.8 \times 19.6 \text{ H'' } 20 \times 20$

$v_f = 20$ m/sec approx.

d. Horizontal final speed = Horizontal initial speed = 20 m/sec

In the absence of a medium, such as air or water, the velocity of a body, falling freely under gravity, will increase in proportion to time according to:

$$v_f = v_f + at$$

Resistive Media

When bodies fall through resisting media, such as air or water, the velocity does not increase indefinitely with time, as the equation of motion for free fall indicates. The falling body will approach a maximum velocity, called the *terminal velocity*. Once this velocity is reached, the falling body no longer accelerates (Fig. 4. 11). The velocity becomes uniform, although the body continues to fall downward under the influence of gravity. For example the terminal velocity of a falling penny is 9 meter/second, and of a feather, or a sheet of paper, is about 0.5 m/s. A skydiver in a spread-eagle position reaches about 58 m/s. But the intrepid diver can increase her terminal velocity by changing the shape of her body, as for example by bringing her legs together and her arms to the side if she wishes to catch up with a partner below flying in a spread-eagle position. A skier speeding down a steep incline will use a similar strategy to reduce the air resistance.

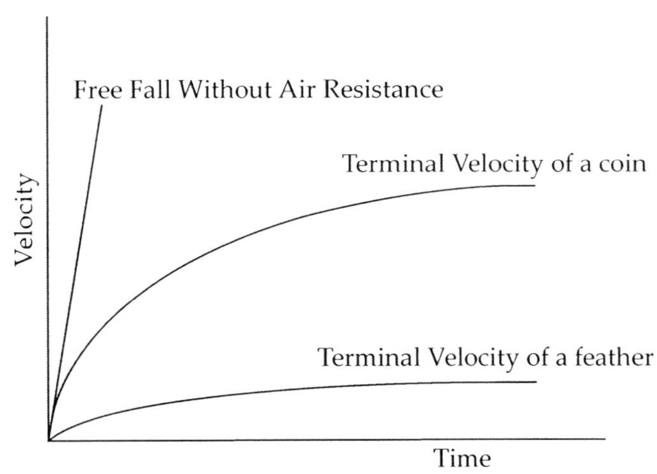

Fig. 4.11: A comparison of the increase of velocity between free fall motion and falling in presence of air resistance.

The reason that the velocity ceases to increase and approaches the terminal uniform velocity is that the force of air resistance increases with the velocity of the falling body. Ultimately, the resistance force becomes equal to the downward force of gravity; after this, the acceleration becomes zero. A quantitative analysis of the situation requires an understanding of Newton's Laws of Motion and their relationship to the concept of momentum; therefore we will return to the derivation in a later chapter. Here we give only the result of the analysis:

$$V_{terminal} = \sqrt{\frac{mg}{ca}}$$

Here m is the mass, g is the acceleration due to gravity, A is the area of the body, and C is a constant that depends on the density of the resisting medium. Thus we see that the terminal velocity does indeed increase with mass, but it decreases with the area of the body and with the density of the resisting medium. Galileo did not derive this result, but he did the crucial spadework to separate the underlying law of motion from the role of the medium.

Galileo wove his ideas about motion into the fabric of a new dynamics. When a body moves with uniform acceleration, as in a free fall or down an inclined plane, the ideal equations of motion are:

$$v_f = v_i + at$$

$$d = v_i t + \frac{1}{2}at^2$$

$$v_f^2 - v_i^2 = 2ad$$

Chapter Questions

1. Cite one specific parallel between the following transformations in European culture and Galileo's approach to understand nature, or in his scientific work:

   recovery of the classics

   challenges to the existing structure

   artists awoke to the wonders of nature

   artists took a realistic approach to depicting nature

   artists developed techniques of linear perspective

   physicians carried out dissections of human cadavers

   the compass was put to practical application for navigation

   new inventions with practical applications, like the printing press

2. Discuss why the Aristotelian professors remained unimpressed when they saw the result of Galileo's experiment that a ball of iron and a ball of wood fall to the ground at the same rate.

3. What underlying law of free fall motion did Galileo discover?

4. Discuss the role of inclined planes, friction and air resistance as applied to the sport of skiing. For example, what factors influence the skier's speed of descent?

5. What were the idealizations that Galileo resorted to for understanding:

   a. motion in free fall?

   b. motion along an inclined plane?

   c. the swinging pendulum

   d. horizontal motion

   e. parabolic projectile motion

6. How do the ideas of Plato and Aristotle come together in Galileo's thinking about motion?

7. On the surface of the moon an astronaut releases a feather and a hammer simultaneously from a height of 2 m. Describe the motion of the two objects.

8. James Bond and a villain are on a plane whose engine fails. The villain jumps out first with the only two parachutes on the plane. Should Bond jump out? How can he catch up with the villain to get the parachute?

9. What is the meaning of terminal velocity?

10. What factors does the terminal velocity of a falling object depend on?

11. Why does a crumpled up ball of paper fall at a faster rate than a sheet of paper?

12. What is the direction of acceleration for a ball falling down from a basketball hoop? What is the direction of acceleration for a ball thrown up to a basketball hoop?

13. A ball is thrown vertically upward. (Ignore air resistance.) At what point during its flight is its acceleration zero?

14. Galileo measured the value of free fall acceleration of a one kg falling rock in Pisa to be 9.8 m/sec^2. What value would his contemporary Shakespeare measure in London?

15. Does Shakespeare have to use a one kg rock?

16. What mathematical patterns did Galileo recognize in the distance covered by a ball rolling down an incline?

17. How did symmetry principles guide Galileo to the radical concepts of eternal horizontal inertial motion with constant velocity, and uniform acceleration under gravity, independent of mass?

18. Suppose you are given two video tapes of a train in ideal motion at constant velocity. The two tapes were started at different times during the passage of the train in front of the cameras. Will you be able to tell which video was taken first? Why or Why not?

19. Do weak-interactions obey time-reversal symmetry?

    Compare the mass of the top-quark with the mass of the up-quark.

    Compare the mass of the muon with the mass of the electron.

20. What is the difference (if any) between Galilean Relativity and

    Einstein's Special Relativity?

21. What were the two main postulates of Einstein's Special Theory of Relativity?

22. What are some of the consequences of Special Relativity for time and space?

23. Is it possible for one person to travel to the future of another person?

24. Is it possible for one person to travel to his own future?

25. Would traveling on a space ship at near the velocity of light increase your life span?

26. Is it possible to travel to the past?

27. What does the distance "one light-year" mean?

28. According to the special theory of relativity, if you measure your own pulse while traveling at very high speeds, you would notice your pulse rate to be faster? slower ?or the same? (As compared to your pulse rate at rest?)

29. If someone on a star 10 million light years away were able to spot any place on earth through a telescope, would he see pyramids in Egypt? Explain your answer.

Math Based Questions

1. Think of some disaster that may occur by using the wrong units for a physical quantity.

2. List 5 distinct units of distance and their conversion factors to meters. List 5 distinct units of time and their conversion factors to seconds.

3. Give one example for a unit of velocity. Give one example for a unit of acceleration.

4. Can you convert one meter distance to time in seconds?

5. During a thunderstorm you see lightning first and hear the thunder later. If the speed that sound travels in air is 340 m/sec, and the delay between the lightning and the thunder is 10 sec, how far away did the lightning originate?

6. You are in a car that accelerates from rest at one meter per second squared. Determine your velocity at the end of 10 seconds.

7. Determine the total distance your car will cover at the end of 10 seconds.

8. You are in a car that accelerates at a constant value (not given). Suppose the distance traveled in the first second is two meters. What is the distance interval that your car will cover in the next one second?

   What is the <u>total</u> distance that your car will cover in four seconds of travel.

9. A driver traveling at 20 km/hour stops the car with his brakes in 4 meters. What is the acceleration of the car? Discuss the sign of the acceleration.

10. A kangaroo can jump straight up in the air about 2.5 meters. What is the speed (in meter/sec) at which it must take off from the ground?

11. If a stone is dropped (not thrown) from a bridge takes 3.7 sec to hit the water, how high is the bridge above the water?

12. A stone is released from a height of 490 m. How long will it take to reach the ground, and with what speed will it strike the ground?

13. A youngster throws a rock horizontally at a speed of 10 m/sec from a bridge 50 m above a river.

    a. How long will it take for the rock to hit the water?

    b. What is the vertical velocity of the rock before it lands?

    c. What is the horizontal velocity of the rock before it lands?

    d. How far from the bridge will the rock hit the water?

14. Suppose a swimmer in the river catches the stone where it touches the water in the above problem. Use symmetry arguments to determine the horizontal and vertical velocity components that the swimmer would have to launch the stone in order to return it to the youngster at the top of the tower.

15. A stone is thrown straight up. After 10 sec have elapsed, it returns to the ground. What height does it reach?

16. Make a table of x and y values to draw a graph of $y = 2x^2$ from $x = 0$ to 5. Draw the graph. By using symmetry, extend your graph to the region $x = 0$ to -5.

17. Apply symmetry to the equation $y = 2x^2$ to discuss the corresponding effects on the graph of the above problem. Carry out the same steps for the function $y = 2x^3$. Make comments on the differences between the two equations.

18. Compare the terminal velocities of a 1 mm diameter raindrop to a 1 mm diameter iron sphere. The density of water is 1000 kg/m^3 and the density of air is 1 kg/m^3. The specific gravity of iron is 10.

19. Carry out a dimensional analysis on the mathematical expressions for distance traveled under uniform acceleration, g.

$$s = v_o t + \frac{1}{2} g t^2$$

$$d = \frac{v_f^2 - v_i^2}{2g}$$

20. Suppose someone derived a formula that did not have the same units on both sides. What would be your conclusion?

21. If L represents the dimensions of length and T the dimensions of time, what are the dimensions of acceleration?

22. Carry out a dimensional analysis of the formula for terminal velocity in the presence of air resistance.

23. You are playing basketball on the deck of a cruise liner moving at a constant speed of 30 km/h. What is the angle to the horizontal at which you should throw the ball to cover the maximum horizontal distance along the deck?

24. Two friends have exactly the same age. Is it possible for one friend to travel to the future of the other? Is it possible for one person to travel to his own future? Is it possible to travel to his own past ?

25. Would traveling on a space ship at near the velocity of light increase your life span?

26. What does the distance "one light-year" mean? Give a quantitative answer.

27. In a particle physics experiment a particle called pion is used to hit a target. The particle pion has a (rest) lifetime of 2.6×10^{-8} second and is accelerated to a speed $0.99c$ with respect to the linear accelerator. A straight beam pipe is used to transport the pions to the target. What would be the maximum length of the beam pipe? What is the length of the beam pipe the pions see?

28. Our galaxy is 10^5 light years across and the most energetic cosmic ray proton are observed with a speed such that they have a relativistic mass of about 10^{10} times the rest mass of a resting proton. Determine how long it would take these energetic protons to traverse our galaxy as measured in the reference frame of: (a) the galaxy and (b) the proton.

Exploration Topics

1. The sites below are places where you can find art, browse, download and print.

   http://www.ibiblio.org/wm/paint/auth/

   http://www.kfki.hu/~arthp/search.html

   http://cgfa.sunsite.dk/fineart.htm

   http://gallery.euroweb.hu/artist.html

   http://www.artcyclopedia.com

   <u>Explore 2 artists each from the following periods</u>

   a. Greek

   b. Medieval

   c. Renaissance

   d. Baroque

   e. Modern

   Chose two works of art from each of these periods that connect to the themes we developed in this course about the connections between science and art. Give the name of the artists and the time period. Explain the connections. Print out the art work. Make sure the work you chose is not referred to in the lecture notes and one that we did not see in class lectures.

2. Drawing from Plato, Aquinas, Bacon, and Galileo, discuss the relative importance and effectiveness of faith, reason, passive observation and controlled experiment in gaining understanding of nature.

Sample Quiz

1. Which of the following choices played a role in Galileo's realization of the pendulum principle?
   a. He paid close attention to the behavior of objects in motion around him.
   b. He quantified the motion with the beat of his pulse.
   c. He carried out experiments with stones and rocks dangling on strings.
   d. All of the above.

2. When a body moves down an inclined plane the <u>total distance</u> traveled
   a. increases uniformly with time.
   b. increases as the square of the time.
   c. is an odd number.
   d. none of the above.

3. What <u>idealizations</u> did Galileo use to arrive at the underlying laws of motion?
   a. He dropped a ball of wood and a ball of iron from a high tower to show that they fall at the same rate.
   b. He separated the role of friction and air resistance from the observed motion.
   c. He used mathematics of the Greeks.
   d. He used an inclined plane to slow down the rate at which bodies fall.
   e. All of the above.

4. Two objects (i) an iron ball 0.5″ in diameter and (ii) a 1/16″ thick metal sheet, one inch square - are dropped from a height of one meter through water in a reservoir. Which one of the following statements best describes the relative motion of the two objects?
   a. They both hit the bottom of the reservoir at nearly the same time.
   b. The brass sheet rises, whereas the metal ball sinks.
   c. The brass sheet falls slowly, like a feather falling through air, while the ball drops fast, like a stone falling through air.
   d. They both hit the bottom of the reservoir at exactly the same time

5. An object is dropped from a height of 10 meters. How long does it take to hit the ground?

   a. 2 sec

   b. 1.43 sec

   c. 1.43 m

   d. 9.8 sec

   e. none of the above

6. You are playing basketball on the deck of a cruise liner moving at a constant speed of 30 km/h. What is the angle to the horizontal at which you should throw the ball to cover the maximum horizontal distance along the deck?

   a. 90 degrees

   b. 45 degrees

   c. 22.5 degrees

   d. need more information

   e. depends on the speed of the ship

7. What was Maxwell's new prediction after unifying electricity and magnetism?

   a. He showed that Newton's theory of gravity needs to be corrected.

   b. He showed that the colors of the rainbow arise from the dispersion of light.

   c. He showed that light must be an electromagnetic wave.

   d. All of the above.

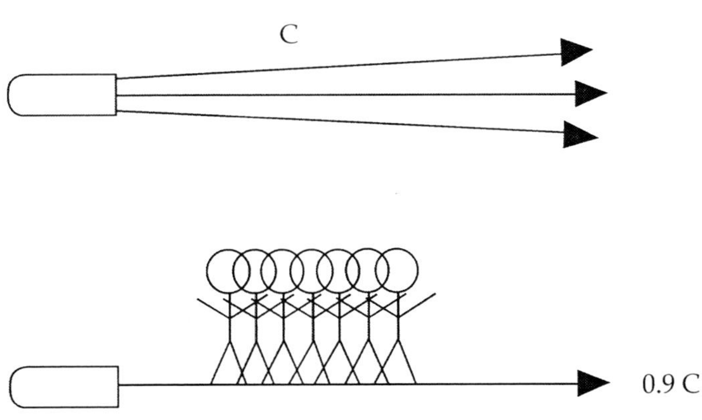

8. In an imaginary (gedanken) situation above, Sam is traveling on a particle beam moving at 0.9 times the speed of light. He observes a parallel beam of light from a laser. What is the speed at which Sam measures the laser beam of light to be traveling?

   a. speed of light, c

   b. zero

   c. 1.9 x the speed of light

   d. none of the above

Answers to Sample Quiz for Chapter 4

1. d
2. b
3. b
4. c
5. b
6. b
7. c
8. a

Homework Questions for Chapter 4 (with sample answers)

1. Cite one specific parallel between the following transformations in European culture and Galileo's approach to understand nature, or in his scientific work: recovery of the classics

   Sample Answer

   * *recovery of the classics:* Galileo had read about the works of Pythagoras and Archimedes that had been translated from Arabic. In particular, Pythagoras' formula that the sum of the first n odd integers is n^2 had played a role in determining the distance that a ball rolls down an inclined plane. As another example, Archimedes' theory of buoyancy influenced Galileo's thinking (and experiments) of objects falling through a medium. Of course, Aristotle's writings had become the standard curriculum by Galileo's time, and must have had influenced Galileo's scientific work tremendously, despite its negative effects.

   challenges to the existing structure

   artists awoke to the wonders of nature

   artists took a realistic approach to depict nature

   artists developed techniques of linear perspective

   physicians carried out dissections of human cadavers

   the compass invention was put to practical application for navigation

2. Discuss the role friction and air resistance as applied to the sport of skiing. For example, what factors influence the skier's speed of descent?

3. What were the idealizations that Galileo resorted to for understanding:

   a. motion in free fall?

   b. motion along an inclined plane?

   c. the swinging pendulum

   d. horizontal motion

   e. parabolic projectile motion

4. How do the ideas of Plato and Aristotle come together in Galileo's thinking about motion?

5. What is the meaning of terminal velocity? What factors does the terminal velocity of a falling object depend on? Why does a crumpled up ball of paper fall at a faster rate than a sheet of paper?

Sample Solution

In free fall, all objects first follow the ideal laws determined by Galileo, so that the velocity increases linearly with time. But when air resistance takes over, the velocity becomes constant after an initial period of acceleration. That final constant velocity is called *terminal velocity*. The factors determining its value are the density of the medium (air or water) the mass of the falling object and the shape of the falling object (which determines the amount of air resistance). That is why a crumbled ball of paper will reach a higher terminal velocity than a flat sheet of the same piece of paper.

6. What is the direction of acceleration for a ball falling down from a basketball hoop? What is the direction of acceleration for a ball thrown up to a basketball hoop? A ball is thrown vertically upward. (Ignore air resistance.) At what point during its flight is its acceleration zero?

7. A driver traveling at 20 km/hour stops the car with his brakes in 4 meters. What is the acceleration of the car? Discuss the sign of the acceleration.

Sample Solution

We are given here the initial velocity (v_i= 20 km/hour) and the final velocity (v_f = 0 km/hour) as well as the distance over which the velocity is changing. We will call the direction the car is traveling the positive direction. Eventually, all our measured quantities will need to be in the same set of units, so we convert km/hour to m/s:

$$\frac{20 \text{ km}}{1 \text{ hours}} \cdot \frac{1000 \text{ m}}{1 \text{ km}} \cdot \frac{1 \text{ hours}}{60 \text{ min}} \cdot \frac{1 \text{ min}}{60 \text{ s}} = 5.\overline{5} \text{ m/s}$$

The kinematic relationship between these quantities is

$$v_f^2 = v_i^2 + 2ad.$$

Isolating the acceleration in this equation we see that

$$a = \frac{v_f^2 - v_i^2}{2d} = \frac{(0 \text{ m/s})^2 - (5.\overline{5} \text{ m/s})^2}{2.4 \text{ m}} = -3.86 \text{m/s}^2$$

The negative sign is correct, as we are going from a large positive velocity to no velocity — our velocity has decreased, indicating a negative acceleration (or a deceleration — "braking").

8. A kangaroo can jump straight up in the air about 2.5 meters. What is the speed (in meter/sec) at which it must take off from the ground?

9. If a stone is dropped (not thrown) from a bridge takes 3.7 sec to hit the water, how high is the bridge above the water?

10. You are in a car that accelerates from rest at one meter per second^2. Determine your velocity at the end of 10 seconds. Determine the total

distance your car will cover at the end of 10 seconds.

11. You are playing basketball on the deck of a cruise liner moving at a constant speed of 30 km/h. What is the angle to the horizontal at which you should throw the ball to cover the maximum horizontal distance along the deck? What is your angle if the ship is moving at twice the speed. Explain your answers.

12. Two friends have exactly the same age. Is it possible for one friend to travel to the future of the other? Is it possible for one person to travel to his own future? Is it possible to travel to his own past ?

13. Would traveling on a space ship at near the velocity of light increase your life span? If you measure your own pulse while traveling at very high speeds, would you notice your pulse rate to be faster? slower ?or the same? as compared to your pulse rate at rest? Explain your answers.

14. What does the distance "one light-year" mean? Give a quantitative answer in km.

15. If someone on a star 10 million light years away were able to spot any place on earth through a telescope, would he see pyramids in Egypt? Explain your answer.

16. Suppose you are on a space ship with a total mass of 1000 kg. You wish to get up to a speed of 0.1c. How much energy in Joules will you need to get up to that speed? At this speed we can use the non-relativistic formula for kinetic energy (K.E. = ½ mc^2). Compare your answer to the appprox total electrical energy consumption in 2012 of 4 Billion kWh. Convert BkWh to Joules. (1 Joule = 1 watt-sec).

Sample Answer

½ mc^2 = 0.5 × 1000 × (3×10^8)^2 = 4.5 ×10^19 Joules

4 BKWh = 4 × 10^9 × 1000 Watt × 60 × 60 sec = 1.44 × 10^16 Joules

You would need more than the entire electrical output of US in a year.

17. Suppose space ship A and space ship B approach each other at a speed of 0.6 c each relative to a stationary observer. What does ship A measure to be the oncoming speed of ship B?

CHAPTER 5

Reading Questions

P. 246: Make a short list of the regular (rhythmic) heavenly motions the ancients observed and recorded.

How did these phenomena represent symmetry and order?

How were these regular motions important to the activities of civilizations?

What aspects were missing in the ancient's examinations of heavenly motions?

P. 247: Make a list of the puzzling aspects of the heavens observed and recorded by ancient civilizations. Why were these puzzling?

How did these aspects show asymmetries and disorder?

What motivated ancient civilizations to keep close track of heavenly motions and phenomena?

P. 248: How did minute irregularities and asymmetries play a role in reaching our present conception of the order of the solar system (planetary system).

What were the practical benefits of an accurate solar calendar?

How did heavenly motions influence cultural practice?

P. 249: What are the special properties of the days of the solstices and the equinoxes?

How did ancient civilizations keep track of these special days?

P. 250: How did these special days influence the cultures of ancient civilizations?

On what days does the sun rise in the exact east?

P. 251: How would you use the orientation of shadows to determine North for a place in the northern hemisphere?

What are the limitations of the sun-dial as a clock?

P. 253:Why did some civilizations adopt a lunar (rather than a solar) calendar?

How did some cultures interpret the dark patches on the moon?

P. 255:What is the relationship between the positions of the full moon and the sun on the horizon?

P. 256:What are the definitions of "opposition" and "conjunction" for the relative position of two heavenly bodies?

In what phase of the lunar cycle does the moon appear in conjunction with the sun?

Why is the moon invisible during the new moon phase?

P. 257:Describe one method to match up the lunar and solar calendars.

P. 258:What were some of the early speculations about the nature of stars?

What was found constant about the appearance and "movement" of the stars?

What were the practical benefits from keeping track of the stars?

P. 259:How would you use the movement of stars through the night to determine the directions east, west, north and south (for a location similar to NY City in the northern hemisphere)?

P. 260:How do the stars help to determine the exact North (at mid-latitudes in the northern hemisphere)?

P. 261:How does the Little Dipper constellation help to keep track of the passing hours at night?

How did Shakespeare describe the nature of stars?

Do you think the shape of the constellations change over a long time?

Do you think stars ever die?

P. 262:How were the constellations used to keep track of the seasons?

P. 264:How were constellations used to guide agricultural activities?

P. 265:How did some of the early civilizations across the globe react (interpret) a total solar eclipse? A lunar eclipse?

P. 268:How do we know that the stars do not disappear from the sky during the day?

P. 269:How did the Babylonians become successful in predicting the occurrence of eclipses?

How do we know that the Milky Way is not just a formation of clouds or mist in the atmosphere?

P. 271: Why did many civilizations treat the appearance of a comet as a sign of impending disaster?

P. 273: How did the coincidence of the first appearance of Sirius in the summer with the rising sun help Egyptian culture?

P. 274: How did we first learn that the year is 365 + 1/4 day long, so that it is necessary to add an extra day to the solar calendar every four years during the leap year?

P. 275: In what direction does the moon travel through the zodiac constellations from night to night?

P. 276: How long does the sun take to make its full circuit through the zodiac constellations?

In what direction does the sun travel through the zodiac constellations ?

If the sun rises in Taurus during the morning, in what constellation will the sun set that same evening?

About how long does the sun spend in each constellation?

P. 280: What distinguishes the appearance and the movements of the planets from the stars?

P. 281: If Venus appears as a morning star, can it appear as an evening star on the same night?

P. 283: Which planets can appear only in conjunction with the sun?

Which planets can appear both in conjunction as well as in opposition to the sun at different times in their cycle through the zodiac?

In which direction do the planets generally move through the zodiac constellations from night to night relative their motion through one night.

What was the long-lasting contribution of Babylonian planetary observations to the future development of astronomy?

P. 285: Why does Mercury appear to move very fast through the zodiac as compared to Saturn which takes 30 years?

P. 287: Describe some of the influences that ancient planetary observations still have on our culture.

P. 288: What is the connection between the tides and the phases of the moon?

P. 290: Describe some of the irregularities the ancient observers found in planetary motions.

What was the long-term significance of these irregularities to future advances in astronomy?

P. 291: Although most of the connections of astrology to human character and destiny are pure fantasy, pick out some aspects of astrology that nonetheless became helpful to the development of astronomy.

P. 293: What is the value to astronomy of detailed star charts such as those made by Herschel?

P. 295: What was the major difference in methodology between the discovery of Uranus and that of Neptune?

P. 296: Make a Table of the number of years it takes the 9 planets (including Pluto) to orbit the sun.

Chapter Questions

1. Describe the changes that take place in the length and the direction of the shadow of an obelisk from the day of the summer solstice to the winter solstice.

2. What is the length of the noon-day shadow on the summer solstice at latitude 23.5 degrees South?

3. In the southern hemisphere, to which direction does the noon-day shadow point?

4. Around sunrise and sunset the shadows are useless to tell time intervals. Why?

5. On what days does the sun rise exactly in the east and set exactly in the west?

6. What is the phase of the moon during (a) a lunar eclipse (b) a solar eclipse?

7. How do we know that the eclipse is not just one of the phases of the moon?

8. If a lunar eclipse occurs when the moon comes between the sun and the earth, why do lunar eclipses NOT occur every month?

9. Why did the ancients think that Mercury is the planet closest to earth? Why did the ancients think that Saturn is the planet farthest from earth?

10. What observations led to the length of the year as 365.25 days?

11. What is the special quality about the north star, Polaris?

12. What is the relationship between the path of the sun, the moon and the planets with the constellations of the zodiac?

13. If the sun sets in the middle of Aries during May 1, in what constellation (approximately) will the sun rise on (a) May 1 (a) June 1 (b) April 1?

14. Name two ways by which the observed motion of the planets differs from the motion of the stars?

11\. Over the course of one week, compile a table of astrological predictions about yourself made by two or more different newspapers. How often do the forecasts agree with (a) each other (b) the future? What is your opinion about horoscopes?

12\. What sign were you born under? What is the astronomical connection between that sign and the location of the sun?

13\. What astronomical influences led us to adopt 12 months of the year, 24 hours in a day and 7 days in the week?

14. How do we know that the stars are still present in the sky during the day?

15. Explain how the stars can be used (a) to keep track of the calendar (b) to keep track of time through the night (c) to keep track of location on earth.

16. Give examples of the influence of astronomy on the life and language of the following religions:

    Egyptians, Babylonians, Greeks, Jews, Christians, Muslims, your own.

17. How do we know that the Milky Way is not a cloud formation?

18. Discuss why the total solar eclipse caused so much consternation among many ancient civilizations.

19. Give two separate reasons why the ancient cultures picked out the planets from among the vast multitude of stars.

20. Discuss two examples to show that the Babylonians were good observers.

21. Among the planets Mars, Mercury and Venus, which one takes the shortest time to make a complete circuit of the zodiac? Arrange the next two in order of increasing time period. What does this order tell us about the locations of the three planets relative to the earth?

22. How long does the moon take to make one complete circuit through the zodiac?

23. How many degrees per hour do the stars revolve around Polaris?

24. For the ancient observers, what was the difference between the appearance and motions of the sun and the moon as compared to the appearance and behavior of the comets and the Milky Way?

25. What led the ancients to believe that there must be a connection between celestial activities and the fate of humans?

26. Why was a lunar calendar more attractive to some cultures than the solar calendar?

27. What are the advantages of a solar calendar over a lunar calendar?

Sample Quiz

1. What happens to the angle between the Pole star and the horizon as you move on the surface of the Earth from the North Pole to the equator

   a. The angle changes from 90 degrees to zero degrees.

   b. The angle changes from 0 degrees to 90 degrees.

   c. There is no change in angle.

   d. The angle changes from 45 degrees to zero degrees.

2. "Nothing is strange, nothing impossible,

   Nor marvelous, since Zeus the father of gods

   Brought night to midday

   What celestial event does the poet refer to?

   a. coming of a comet

   b. lunar eclipse

   c. the milky way

   d. a total solar eclipse

3. Why did the ancients think that Mercury is the planet closest to earth and that Saturn is the farthest from earth?

   a. Because Mercury always appears in conjunction with the sun, but Saturn can appear either in conjunction or in opposition.

   b. Because Mercury is the god of messengers while Saturn is the god of time.

   c. Because Mercury appears after sunset while Saturn appears before sunrise.

   e. Because Mercury moves the fastest through the zodiac while Saturn moves the slowest.

4. What scientific aspect distinguishes the constellations of the zodiac from other constellations?

   a. They help to predict the future.

   b. They appear to lie in the path of the Sun, the Moon and the Planets.

   c. They move from east to west.

   d. The stars in these constellations maintain a fixed pattern.

5. On which of the following days is the noon shadow cast by a pole the shortest in length (in the northern hemisphere)?

    a. summer solstice

    b. spring equinox

    c. winter solstice

    d. autumnal equinox.

6. What distinguishes the planets from the stars?

    a. They inspired painters and poets.

    b. They move with respect the constellations.

    c. They move among the constellations of the zodiac.

    d. none of the above

    e. (b) and (c).

7. Which of the following takes the longest time to orbit around the zodiac?

    a Jupiter | b. Saturn

    c. Mars | d. Mercury

8. At approximately what time of the day does the full moon *set*?

    a. at mid-day | b. at sunrise

    c. at sunset | d. none of the above

9. What is the *evidence* that the Milky Way is not a cloud formation?

    a Because it always goes through the same set of constellations.

    b Because it is never visible on cloudy nights.

    c Because we live in the Milky Way Galaxy.

    d Because we call it our galaxy.

    e All of the above.

Answers to Sample Quiz for Chapter 5

1. a
2. d
3. e
4. b
5. a
6. e
7. b
8. b
9. a

Homework Questions for Chapter 5 (with sample answers)

1. What are the special properties of the following days: summer solstice, winter solstice, the autumnal equinox, the vernal equinox?

2. What is the length of the noon-day shadow on the summer solstice at latitude 23.5 degrees South? In the southern hemisphere, to which direction does the noon-day shadow point?

   Sample Solution

   On summer solstice (summer for people on the southern hemisphere, which is winter for us, around Dec. 22), the Sun at midday is directly overhead at latitude 23.5 degrees South. Therefore no shadow is casted. (The situation is exactly what happened at Syenne that Erastothenes had noticed.)

3. If the sun sets in the middle of Aries during May 1, in what constellation (approximately) will the sun rise on (a) May 1 (a) June 1 (b) April 1?

4. What observations led to the length of the year as 365.25 days?

5. In what direction does the moon appear to travel through the zodiac constellations from night to night? In what direction does the moon appear to travel across the sky in one night?

6. What is the phase of the moon during (a) a lunar eclipse (b) a solar eclipse? How do we know that the eclipse is not just one of the phases of the moon? If a lunar eclipse occurs when the moon comes between the sun and the earth, why do lunar eclipses NOT occur every month?

7. How do we know that the stars are still present in the sky during the day? Explain how the stars can be used (a) to keep track of the calendar (b) to keep track of time through the night (c) to keep track of location on earth. What are the special qualities about the north star, Polaris?

8. How did the ancients arrive at a scheme to organize the distance of the 7 heavenly objects relative to a central earth? Earth, Moon, Mercury, Venus, Sun, Mars, Jupiter, Saturn?

9. Explain the meaning of conjunction and opposition relative to the bodies (Moon, Sun), (Sun, Venus), (Sun, Mars). Which case of opposition never happens?

10. Explain how the stars can be used (a) to keep track of the calendar (b) to keep trackof time through the night (c) to keep track of location on earth. What are the special qualities about the north star, Polaris?

11. How did ancient cultures picked out the planets from among the vast multitude of stars.

Sample Answer

The five planets visible to the naked eye are different from the vast multitude of stars in the constellations because they move (hence planet meaning "wanderer") against the background of constellation stars. The constellation stars themselves don't appear to move from night to night, nor from year to year. Only over millennia is there noticeable change (due to the so-called proper motion) to a handful of stars.

12. What is the relationship between the path of the sun, the moon and the planets with the constellations of the zodiac?

13. Among the planets Mars, Mercury and Venus, which one takes the shortest time to make a complete circuit of the zodiac? Arrange the next two in order of increasing time period.

    What does this order tell us about the locations of the three planets relative to the earth?

14. For the ancient observers, what was the difference between the appearance and motions of the sun and the moon as compared to the appearance and behavior of the comets?

15. How do we know that the Milky Way is not a cloud formation?

CHAPTER 6

Reading Questions

P. 299: How was the Greek approach to astronomy different from the Egyptian and Babylonian approaches?

P. 300: According to the Greek model of the celestial sphere, why does the North star remain motionless?

What motivated Pythagoras to assign separate spheres (from the celestial sphere) to the sun, moon and each of the planets?

What is retrograde motion?

P. 304: Why did most of the Greek thinkers reject the idea of a spinning earth as proposed by some of the Greek philosophers?

P. 305: What is the phase of the moon when it is rising above the horizon at the same time that the sun is setting?

What is the phase of the moon when it is rising above the horizon at the same time that the sun is rising?

P. 306: If the tide is at its maximum height at 9 am on Monday, at what time on Tuesday (the next day) will the tide be at its highest?

What is the phase of the moon during a total lunar (solar) eclipse?

P. 309: What led Plato to believe that despite all the apparent irregularities observed in planetary motion science would eventually find the underlying regularity in those paths?

P. 310: How was Eudoxus approach to understanding the heavens distinct from Plato's?

P. 311: What is the reason for the seasons according to Eudoxus' model?

P. 313:How did Eudoxus know that the moon's path through the zodiac is tilted by about 5° relative to the sun's path (ecliptic)?

What are the necessary conditions for the moon's phase and the moon's path necessary for a lunar eclipse to occur?

P. 315:What are the parallels between Eudoxus' new approach to the heavens and those of the artists in his time?

What are the regularities observed in the brightness of Mars's appearance?

What is the location of Mars on the horizon relative the sun when it is in retrograde motion?

Did Eudoxus's work make the heavens' descriptions simpler than Pythagoras' model?

P. 316:How was Eudoxus' model an improvement over Pythagoras's model?

P. 317:What is the relationship between the latitude on earth and the altitude of the Pole star?

P. 318:Why did Aristotle place comets in the earth's atmosphere rather than in the heavens?

P. 319: What were the lasting contributions of the early Greek astronomers and thinkers?

P. 321: What was Aristarchus' explanation for the fixed position of the Pole star?

How was Aristarchus' system a simplification over the Eudoxus/Aristotle's system of heavenly spheres?

P. 322:Why do the sun and the moon appear to be about the same size (or subtend the same angles to the eye)?

P. 323:Does the moon take the same number of days and hours to change from new moon to half-moon as it takes to go from half-moon to full-moon?

How did Aristarchus conclude that the size of the earth's shadow on the moon is twice the moon's diameter?

P. 325:Should we dismiss Aristarchus' results due to his large errors?

Why did his contemporaries and following astronomers reject Aristarchus' system?

P. 326:What was Aristarchus' possible explanation for the absence of parallax (absence of changing angles between stars) between summer and winter?

How could Aristarchus have offered a more convincing case for his system?

P. 327:How did Apollonius explain retrograde motion?

P. 328:How was Apollonius earth-centric system different from Aristotle's system?

P. 329:What was Apollonius' lasting contribution to the development of science?

P. 330:Why was making an accurate and detailed star map (by Hipparchus) important to the further progress of astronomy?

What new feature did Hipparchus add to his observations of the stars and his star-map?

P. 331:How did Hipparchus advance his observational techniques to achieve a star map accurate to 10 arc-minutes?

How many arc-minutes are there in one degree of arc?

What new discovery did Hipparchus make due to the higher level of precision in observations?

P. 333:What is the impact of the precession of the equinox on your astrological sign?

P. 334:What was Hipparchus' explanation for the precession of the spring equinox?

P. 336:Why did Ptolemy have to add more eccentrics and circles to Hipparchus/ Apollonius' model?

Despite the complexity what were the benefits of Ptolemy's model?

P. 337:Why did Ptolemy's model drift so far from simplicity?

P. 338:How were Ptolemy's estimates for heavenly distances an advance over Aristarchus results?

P. 339:What new observations did Ptolemy record besides stars and planets?

P. 340:How do navigators on long ocean journeys determine the longitude of their position?

P. 341:Why were Arabs generally interested in astronomy?

P. 342:How did Islamic astronomers advance measurement techniques over the Alexandrian astronomers?

P. 344:Compare the size and accuracy of the space telescope HIPPARCOS's star catalog with that of Hipparchus' star catalog.

P. 355 Why was Messier interested in cataloguing diffuse (nebular) objects rather than stars?

P. 346:What is the one of the principal differences between the goals of the SDSS project and the HIPPARCOS project?

P. 347:How did the different units of distance measurement change with the increasing distances of stellar objects?

P. 349:What is a quasi-stellar object (Quasar)?

P. 350: What major discovery became possible due to Bessel's accurate star catalog of 50,000 stars?

P. 351: How big was the parallax angle to the first star which was found to show the parallax shift due to the earth's revolution around the sun?

What is the farthest star-distance that the HIPPARCOS satellite could measure using the trigonometric (parallax) method?

P. 352: Even though their distance estimate to Sirius was excellent for the first luminosity-based measurement, what assumption did Newton and Huygens make that gave them the distance to Sirius that was off by a factor of 2?

How did it become possible to first compare the intrinsic luminosity of stars?

P. 353: How did Leavitt know that the Cepheid variables she was studying were all at the about the same distance from earth?

What is the period-luminosity relationship that Leavitt discovered?

P. 354: Why was it not possible to use the parallax method to determine the distance to the Cepheid variables?

How did Hubble determine the distance to the Andromeda nebula (galaxy)?

What assumption did Hubble make about stars to determine distances to farther out galaxies in which he did not find Cepheid variables?

Math Based Topics

Aristarchus' Analysis for Distance to the Sun and Moon

Once a month, when exactly half of the moon appears illuminated, the earth-moon line forms a right angle with the moon-sun line (Fig. 6.1a). The angle between the earth-moon and the earth-sun lines can be measured. Aristarchus determined 87 degrees. In any right triangle, where one angle is 87 degrees, the ratio of the short adjacent side to long hypotenuse is 1:19. In this case, the short side is the earth-moon line, and the hypotenuse is the earth-sun line. Using simple triangle geometry, Aristarchus concluded that the sun is 19 times farther away from the earth than the moon is. Actually, modern results show that this angle is not 87 degrees, but 89 degrees and 50 minutes, so close to 90 degrees that only precise telescopes can measure the difference from 90 degrees. Therefore the modern value of the ratio of the earth-sun to earth-moon distance is not 19:1, but 395:1.

To determine the size of the sun relative to the moon, Aristarchus applied his geometrical analysis on the occasion of a total eclipse of the sun. He observed that the disk of the moon almost covers the disc of the sun to bring the eerie twilight (Fig. 6.1b). By similar triangle analysis, the ratio of the sizes is equal to the ratio of the distances. Therefore the sun is 19 times larger than the moon. It appears to be nearly the same size as the moon because it is also much farther away from us than the moon is.

He had yet to devise a method to compare the relative sizes of the sun and the earth. Now he factored in additional observations made during a lunar eclipse. When the earth comes between the sun and the moon, the moon enters the earth's shadow, taking a certain time (t) to become eclipsed (Fig. 6.1c). Then it stays in the earth's shadow for some time, which when measured, turns out to be once again nearly the same as the first time interval, t. From these careful observations, Aristarchus concluded that the width of the earth's shadow must be twice the diameter of the moon.

Again, by two consecutive applications of similar triangle geometry, as explained below, he concluded that the size of the sun is about 7 times that of the earth (Fig. 6.2). Modern measurements show that the width of the earth's shadow is 2.5 times the lunar diameter. Besides, Aristarchus' value of 1/19 for the ratio: (diameter of the moon/diameter of the sun) was also in error. Today we know that the diameter of the sun is 109 times larger than that of the earth.

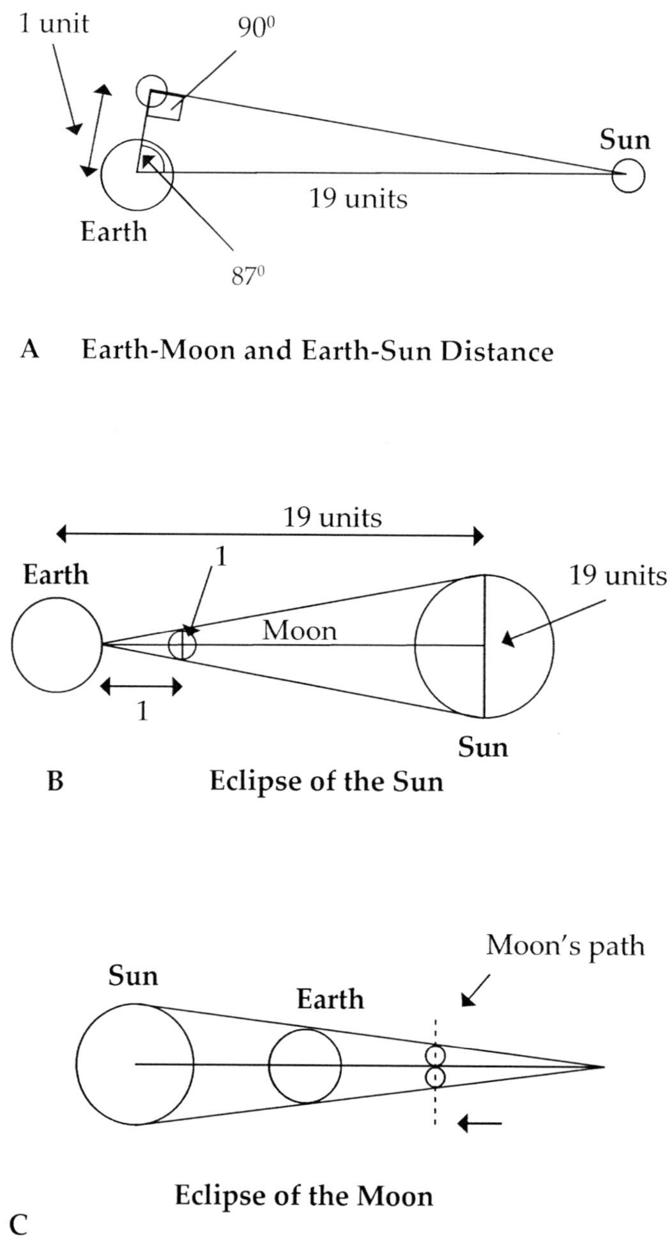

Fig. 6.1: Aristarchus applies triangle geometry to the heavens (a) during the night, when exactly half of the moon is illuminated, (b) during the total eclipse of the sun, and (c) during the eclipse of the moon.

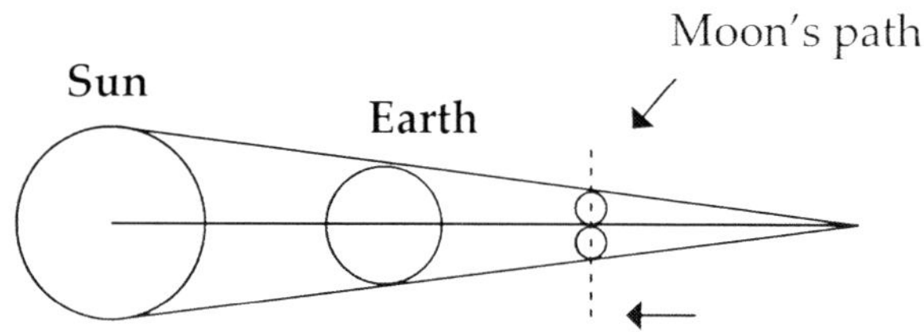

Eclipse of the Moon

C

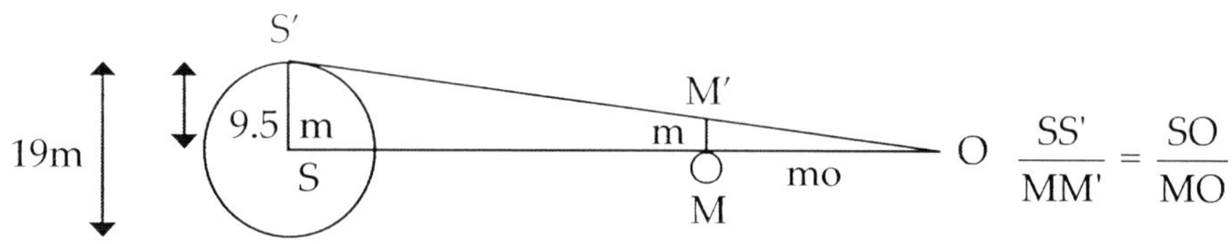

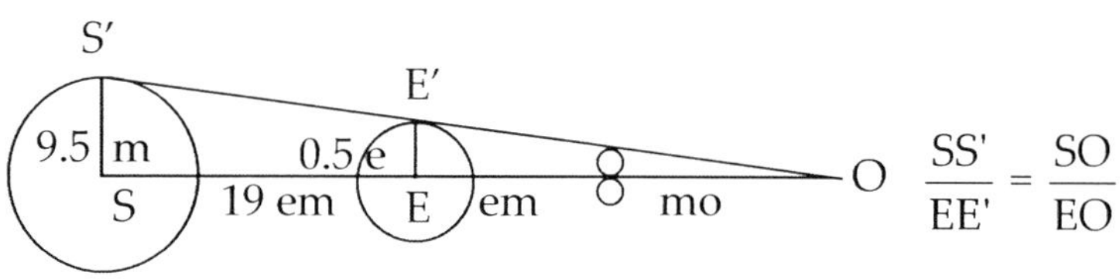

By similar

$$\frac{SS'}{EE'}=\frac{9.5m}{m}=\frac{19em+em+mo}{mo}=\frac{SO}{MO}$$

$$9.5=\frac{20em+mo}{mo}$$

$$9.5mo = 20em\text{-}mo$$

$$8.5mo = 20em$$

$$mo = 2.35em$$

By similar

$$\frac{SS'}{EE'}=\frac{9.5m}{0.5}=\frac{20em+mo}{em+mo}\frac{SO}{MO}=\frac{20em+2.35em}{em+2.35em}=\frac{22.35em}{3.35em}=6.67$$

$$\frac{s}{e}=6.67$$

Fig. 6.2: Aristarchus' method to determine the size of the sun relative to the size of the earth.

Elementary Trigonometry

(a)

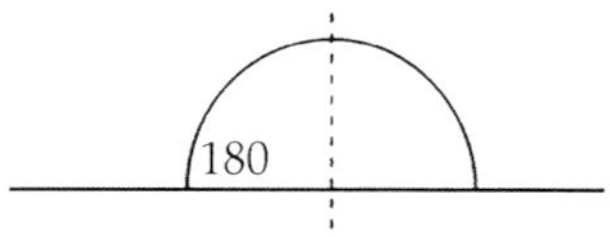

$2\ \pi$ radians = 360 degrees

$$\text{Angle (radians)} = \frac{\text{arc}}{\text{radius}}$$

(b)

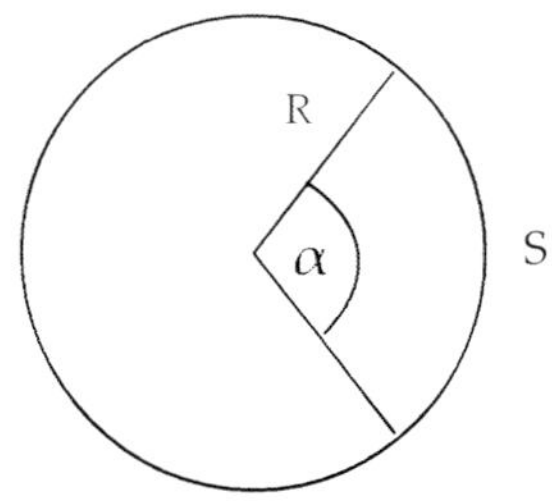

$$\alpha = \frac{S}{R}$$

(c)

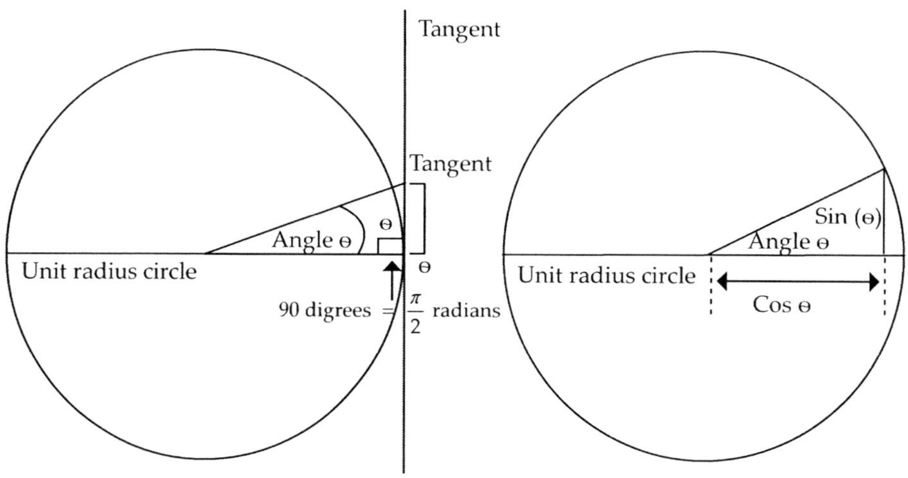

$$\text{angle (radiuns)} = \frac{\text{arc}}{\text{radius}} = \frac{\theta}{1} = \theta$$

$$\text{tangent}(\theta) = \frac{\text{Vertical Side}}{\text{Horizontial Side}} = \frac{\approx\theta}{1}$$

$$\theta \approx \text{tangent }(\theta)$$

(d)

Right Angle Triangle

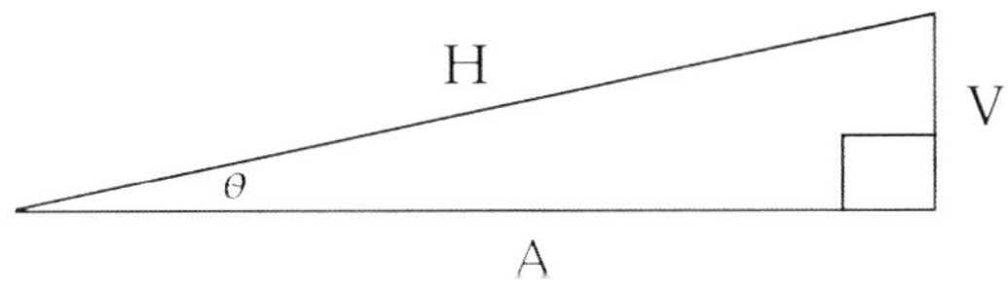

$$\sin(\theta) = \frac{V}{H} \qquad \cos(\theta) = \frac{A}{H} \qquad \tan(\theta) = \frac{V}{A}$$

For small θ

$$\theta\ (in) \approx \frac{V}{A}$$

Fig. 6.3: Elementary angular and trigonometric relations. (a) The angle defined by a straight line is 180 degrees.(b) The angle in radians is defined by the ratio of the arc length (s) to the radius (R). (c) The definition of the tangent of the angle θ as the distance along the tangent between the two radii that include the angle θ. The radius of the circle is of unit length. The length of the tangent segment is approximately equal to the length of the arc θ, which leads to the very useful approximation $\theta \approx \tan\theta$, for small θ. Note also the important relationships between the angle in radians and the definition in degrees. On the right is the definition of sine and cosine of the angle, using a circle of unit radius. (d) Definitions of sine, cosine and tangent based on the legs of a right angle triangle.

Example

On the face of a clock, what is the angle in degrees, and in radians that a minute hand will sweep in one second of time?

When the minute hand goes around one complete revolution it travels 360 degrees of angle to mark 60 seconds of time.

In one second of time it will travel 6 degrees of angle.

180 degrees = π radians

In one second it will travel 6 degrees = 6 degrees

$$= \frac{\pi \text{ radias}}{180 \text{ digrees}} = 0.104 \text{ radians}$$

Example

Stick out your thumb and extend your arm. What is the angle in arc-minutes spanned by the thickness of your thumb (≈1 cm) at the distance of your arm (≈1 cm)?

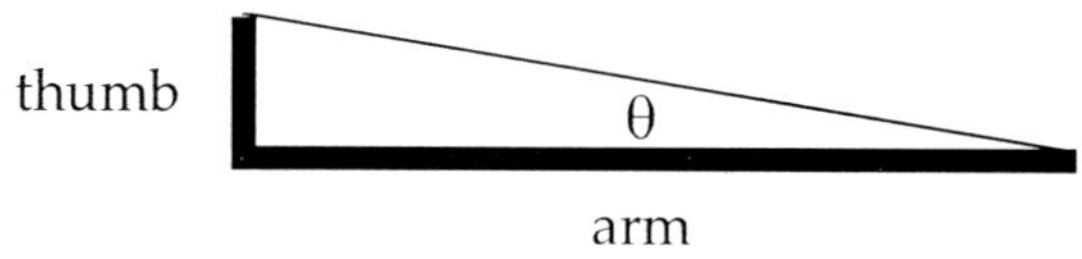

$$\tan\theta = \frac{\text{thumb}}{\text{arm}} \approx \theta = \frac{1}{100} = 0.01\,\frac{180 \text{ degrees}}{\pi \text{ radians}} = 0.57 \text{ degrees} = 34 \text{ minutes}$$

Fig. 6.4: The angle subtended to the eye by the width of the thumb at arms length.

Example

If the moon's face is 30 minutes in diameter, what is the absolute distance of the moon from the earth. Use Eratosthenes diameter of earth, and Aristarchus value for relative size of the earth and the moon. Compare your answers to modern values.

First, we calculate the relative diameter of the moon to the diameter of the earth.

Aristarchus determined that $\dfrac{\text{diameter of sun}}{\text{diameter of earth}} = 6.7$and

$$\frac{\text{diameter of sun}}{\text{diameter of earth}} = 19$$

$$\text{Therefore } \frac{\dfrac{\text{diameter of sun}}{\text{diameter of earth}}}{\dfrac{\text{diameter of sun}}{\text{diameter of moon}}} = \frac{\text{diameter of moon}}{\text{diameter of earth}} = \frac{6.7}{19} = 0.35$$

Next we determine the numerical value for the radius of the moon from the value of the radius of the earth.

Eratosthenes determined that the circumference of the earth is 25,000 miles.

Convert to km, using 1 mile = 1.61 km

Circumference of earth = 40,250 km

Circumference of a circle is $2\pi R$

$2\pi R = 40{,}250 \Rightarrow R = 6405$ km

(The modern value for the radius of the earth is 6378 km)

Diameter of earth = 12810 km

From Artistarchus ratio, diameter of moon = $0.35 \times 12810 = 4483$ km

(The modern value is 3476 km)

The radius of the moon (RM) = 2241 km

Finally, we can calculate the earth-moon distance from the angular size of the moon.

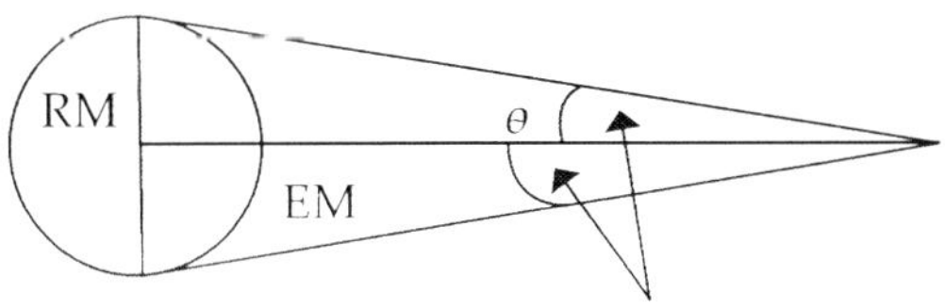

$$\frac{RM}{EM} = \tan\theta \approx \tan\theta \approx \theta \text{ in radians}$$

Observations show that the half angle subtended by the moon's face is 15 minutes.

$$15 \text{ min} = \frac{15}{60} \text{ degrees} = 0.25 \text{ degrees } \frac{\pi \text{ radians}}{180 \text{ digrees}} = 0.00436 \text{ radians}$$

$$\frac{RM}{EM} = \tan\theta \approx \theta \text{ in radians}$$

$$EM = \frac{RM}{\theta} = \frac{2241}{0.00436} = 513990 \text{ km}$$

The earth-moon (EM) distance is 513990 km

The modern value for the earth-moon distance is 384,500 km.

We can also estimate the earth-sun distance from Aristarchus' estimate that the sun is 19 times further from us than the moon.

Earth-sun distance = 19 × 513990 = 9765810 km = 9.77 × 10^6 km

The modern value is 149 × 10^6 km.

Although the results are substantially off the presently known values, they are of the right order of magnitude. The discrepancies should not be too surprising, since we do know that Aristarchus had compounded several errors in his original measurements.

Parallax

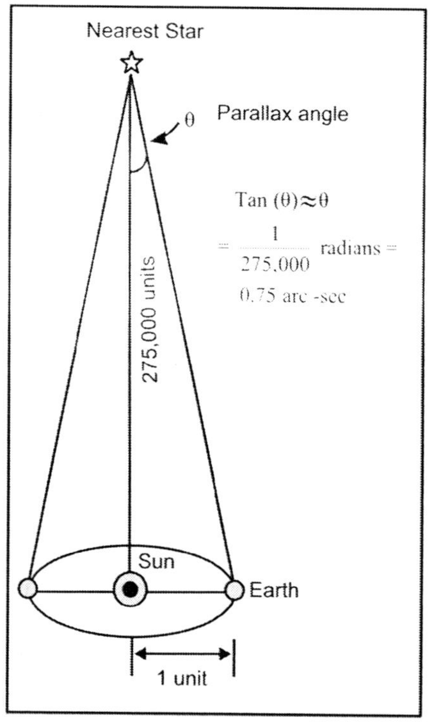

Fig. 6.5: The angle of parallax to the nearest star is exceedingly small and difficult to measure, even with a telescope. Later generation telescopes made it possible to measure angles of arc-second.

One Astronomical Unit = distance from sun to earth = 1.5×10^{11} meters

Chapter Questions

1. What is the phase of the moon during (a) a lunar eclipse (b) a solar eclipse?
2. How do we know that the eclipse is not just one of the phases of the moon?
3. Why are the days longer in the summer than in the winter?
4. Why was it so important to determine the exact angle of the ecliptic?
5. Why do eclipses not occur every month?
6. Why did Aristotle believe that comets do not belong to the heavens? Give both his theoretical reasons as well as his observations.
7. What were the philosophical compulsions that drove the early thinkers to insist that planetary motion must be composed of circles?
8. What is retrograde motion? What impact did it have on the development of scientific thought?
9. Why did the Greeks start with the model that the earth must be at the center of the universe?
10. What astronomical observations continued to defy the earth-centric model through the centuries?
11. How did the earth-centric thinkers modify their theories to continue to maintain their views?
12. Why could the variations in the brightness of the planets not be simply attributed to atmospheric effects?
13. What led Aristarchus to propose a heliocentric model for the heavens?
14. What are the similarities between Aristarchus' approach to scale the heavens and Eratosthenes' approach to scale the earth?
15. During a total solar eclipse, the moon hides the entire sun. What two important scales about the moon and the sun does this phenomenon reveal?
16. What were the arguments against the heliocentric model of planetary motion when it was first presented by Aristarchus?
17. How did the existence of a detail star map advance astronomy in the time of the Alexandrians?
18. Trace the development of increasing complexity in the description of planetary motion from Pythagoras to Ptolemy.
19. Identify the orientation of a plane intersecting a cone to form an ellipse?
20. What were the factors that contributed to Hipparchus' many astronomical successes? Do you think he was just lucky?

21. Why can we see Sirius in the northern hemisphere even if it is below the celestial equator? Explain with the help of a diagram.

22. Discuss how the introduction of a new observational method can lead to a major advance in science.

23. Discuss how the introduction of a new analytic method can lead to a major advance in science.

24. Why did Hipparchus bother to make a catalog of the stars?

25. Name two instruments Hipparchus used to improve precision of measurements of the location of stars.

26. What attribute of the stars did Hipparchus catalog, besides their celestial co-ordinates.

27. "When the moon is in the seventh house and Jupiter aligned with Mars,

    Then peace will guide the planets and Love will fill the air.

    This is the dawning of the age of Aquarius."

    Find three separate astronomical (not astrological) statements in the above song.

    In what constellation will the sun rise on the spring equinox during the year 6000 AD?

28. How large was Ptolemy's estimate for the entire universe?

29. What is the distance to the nearest star (from modern estimates)?

30. What is the name of the astronomical measurement instrument developed to a high state by the Islamic astronomers? What were the capabilities of this instrument?

31. What is the distance to one of the nearest galaxies; name the galaxy you refer to?

32. How large is the universe according to present scientific thinking?

33. Ptolemy was the first to take notice of nebulae. What are nebulae?

34. Make a list of the techniques used to scale the cosmic ladder starting from attempts to determine the size of the earth to attempts to determine the size of the universe.

34. By what factor did satellite Hipparcos accuracy of angle resolution improve over astronomer Hipparchus accuracy?

35. What benefits resulted from astronomer's efforts to make detailed catalogs and star maps?

36. What benefits do we hope will result from modern astronomer's efforts to make detail maps of galaxies?

37. What is the distance to the nearest galaxy?

Math Based Questions

1. How many degrees per hour do the stars revolve around Polaris?
2. Convert 1 second of arc to fractions of a degree.
3. On the face of a clock, what is the angle (in degrees) that a second hand will sweep in 15 seconds?
4. Calculate the angle represented by the thickness of a 1/4" thick pencil at 15 ft from the eye. Give your answer in radians and in degrees. Be careful to convert units as needed.
5. Use similar triangle diagrams to show that, if the sun is 19 times farther from the earth than the moon is, the sun is also 19 times larger in diameter than the moon.
6. What is the ratio of the size of the moon to the size of the earth. Using Eratosthenes estimate of the size of the earth, give the size of the moon. Give the size of the sun, also from Eratosthenes' earth size.
7. How many earths is it possible to fit inside the *volume* of the sun?
8. Compare the surface area of the sun to that of the earth.
9. How many degrees of longitude do you have to travel before there is a time change of 1 hour?
10. Ptolemy estimated the distance to the celestial sphere to be 20,000 earth radii, and distance between earth and sun to be 1210 earth radii. If Aristarchus was right about the heliocentric system, estimate the expected angle (**in degrees**) of parallax to a star on the celestial sphere due to the different positions of the earth between summer and winter solstice. See Diagram below. Choose the closest answer and consider why Ptolemy rejected Aristarchus. Would Hipparchus be able to measure such an angle?

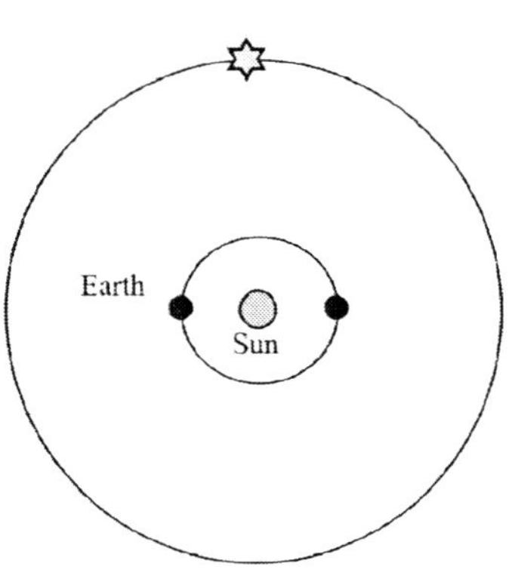

11. How many times brighter is a star of apparent magnitude 1 as compared to a star with apparent magnitude 6?

12. Precession of the equinox takes 26,000 years for a 360 degree full circuit. How many years would it take Hipparchus with his accuracy of 10 arc minutes to detect a shift in the position of the sunrise among the constellations?

13. How many minutes are there in one year? If it takes 26,000 years for the equinotical position of the sun to make a complete circuit of the heavens, calculate the difference in the time of the vernal equinotical sunrise between one year and the next. This is the amount of the precession of the equinoxes in one year. The movement of the equinoxes affects the relationship between the length of the year measured as the return of the sun to a particular star (the tropical year) and the year as measured by the return of the sun to one of the equinotical points (sidereal year). By how much is the tropical year shorter than the sidereal year?

14. Draw diagrams to show how the heliocentric model was successful in explaining the variations in the brightness of the planets, and why an earth-centric system could not explain this observation?

15. Determine the number of sidereal months and the number of synodic months which will give an eclipse super-cycle of 18.03 years.

16. What is the meaning of one astronomical unit?

    Give your answer in terms of km, earth radii, and the time it would take light to travel that distance.

Exploration Topics

1. Cite and discuss cases where observation alone was not enough to arrive at understanding. It was important to use reason and insight. Choose examples from motion on earth and from motion in the heavens.

2. Give examples of the influence of astronomy on the life and language of the following civilizations:

   Egyptians, Babylonians, Greeks, Muslims, Christians, our own.

3. When Pythagoras proposed that the earth must be a sphere, he faced many counter arguments from his contemporaries. Speculate as to what these objections to a round earth may have been? Present a parallel debate over the motion of the earth at the time of Aristarchus.

4. Discuss the developments in this chapter in terms of the quote at the beginning of this chapter.

5. Comment on the following statement, giving examples from the development of scientific thought:

   "Inconsistencies are often the seeds of important creative achievements."

6. Aristotle argued against an infinite universe because it would have no center? Discuss this argument and the impact that an infinite universe would have on his cosmology.

7. Comment upon Aristotle's methods and his cosmology using the following passage from On the Heavens:

   "The truth of it is also clear from the evidence of the senses, enough at least to warrant the assent of human faith; for throughout all past time, according to the records handed down from generation to generation, we find no trace of change either in the whole of the outermost heaven or in any one of its proper parts. It seems too that the name of this first body has been passed down to the present by the ancients...Thus they, believing that the primary body was something different from earth and fire and air and water, gave the name *aether* to the uppermost region, choosing its title from the fact that it "runs always" and eternally."

9. The Hubble space telescope shows a close up view of one of the nebulae Ptolemy spotted. What does the Hubble telescope show?

Sample Quiz

1. What are the necessary conditions to have a total solar eclipse?

   a. The moon must be in full phase and in opposition to the sun.

   b. The moon must in new phase and its position should intersect the ecliptic.

   c. The moon must be in new phase and in conjunction with the sun.

   d. The moon must be in full phase and its position should intersect the sun's path.

2. Which of the following statements is true?

   a. The time interval between the new moon and the half moon is the same as the time interval between the half moon and the full moon.

   b. The time interval between the new moon and the half moon is more than the time interval between the half moon and the full moon.

   c. The time interval between the new moon and the half moon is less than the time interval between the half moon and the full moon.

   d. None of the above.

3. Precession of the equinox takes 26,000 years for a 360 degree full circuit. How many years would it take Hipparchus to detect a 10 arc-minute shift in the position of the vernal equinox sunrise among the zodiac constellations. Chose the closest answer.

   a. 1064 b. 100

   c. 72 d. 12

4. If the speed of light is 3×10^8 m/sec, express the distance of one LY in km.

   a. 10^{13} km b. 10^{16} km c. 10^{17} km

   d. 10^{11} km e. 10^{9} km

5. Convert Ptolemy's estimate of the distance to the celestial sphere of 22,000 earth radii into LY. Radius of earth = 6400 km. By how much was Ptolemy off, if you compare the currently known distance to the nearest star in LY. Chose the closest answer.

   a. He was close to the modern distance to the nearest star.

   b. His distance estimate was 100 times too small.

   c. His estimate was thousand times too small.

d. His estimate was million times too small.

e. His estimate was a billion times too small.

6. Calculate the angle represented by the thickness of a 1/4″ thick pencil at 15 ft from the eye. Give your answer in arc-degrees, arc-minutes or arc-seconds whichever is most appropriate; chose the closest answer.

a. 5 arc-minutes

b. 0.00139 degrees

c. 5 degrees

d one degree

e. 0.5 arc-seconds

7. Why could the variations in the brightness of the planets not be simply attributed to atmospheric effects?

a. Because the brightness of the planets depends on the phase of the moon.

b. Because the brightness depends on the position of the planets in the zodiac.

c. Because the brightness occurs on a regular basis.

d. Because the brightness depends on the location of the planet relative to the location of the rising (or setting) sun.

e. All of the above.

8. What was the common theme that ran through the models of all the Greek and Alexandrian astronomers (from Pythagoras to Ptolemy)?

a. They all used circles.

b. They all put the earth at the center of motion.

c. They all used more than one circle for each heavenly body.

d. All of the above.

Answers to Sample Quiz for Chapter 6

1. b
2. c
3. d
4. a
5. d
6. a
7. c
8. a

Homework Questions for Chapter 6 (with sample answers)

Give One or Two Sentence Answers

1. What are the definitions of "opposition" and "conjunction" for the relative position of two heavenly bodies? Which planets appear only in conjunction with the sun?

   Sample Answer

   Opposition means planet appears on the opposite side of the horizon as the sun sets or rises Conjuction means same side

   Venus and mercury.

2. How was the Greek approach to astronomy different from the Egyptian and Babylonian approaches?

   Sample Answer

   The Greeks tried to develop models based on spheres and circles while the others recorded astronomy facts. It is the difference between trying to gain understanding and gaining knowledge.

3. According to the Greek model of the celestial sphere, why does the North star remain motionless?

4. What motivated Pythagoras to assign separate spheres (from the celestial sphere) to the sun, moon and each of the planets?

5. What is retrograde motion? What impact did it have on the development of scientific thought?

6. Why did most of the Greek thinkers reject the idea of a spinning earth as proposed by some of the Greek philosophers?

7. How was Eudoxus approach to understanding the heavens distinct from Plato's?

8. What is the location of Mars on the horizon relative the sun when it is in retrograde motion?

9. How was Eudoxus' model an improvement over Pythagoras's model?

10. What is the relationship between the latitude on earth and the altitude of the Pole star?

11. Why did Aristotle place comets in the earth's atmosphere rather than in the heavens?

12. Why do the sun and the moon appear to be about the same size (or subtend the same angles to the eye)?

13. Does the moon take the same number of days and hours to change from new moon to half-moon as it takes to go from half-moon to full-moon? Explain your answer.

14. How did Aristarchus conclude that the size of the earth's shadow on the moon is twice the moon's diameter?

Sample Answer

Because the moon stayed in total eclipse for about twice as long as the time it took the moon to enter from full to total eclipse.

15. What was Aristarchus' possible explanation for the absence of parallax (absence of changing angles between stars) between summer and winter?

16. How did Apollonius explain retrograde motion?

17. What was Apollonius' lasting contribution to the development of science?

18. Why was making an accurate and detailed star map (by Hipparchus) important to the further progress of astronomy?

19. What new feature did Hipparchus add to his observations of the stars and his star-map?

20. How did Hipparchus advance his observational techniques to achieve a star map accurate to 10 arc-minutes?

21. What new discovery did Hipparchus make due to the higher level of precision in observations?

22. During a total solar eclipse, the moon hides the entire sun. What two important scales about the moon and the sun does this phenomenon reveal?

23. How large was Ptolemy's estimate for the entire universe?

CHAPTER 7

Reading Questions

P. 356: Using Whitehead's quote at the beginning of this chapter, name one pair each of astronomers from the following periods: Alexandrian, Renaissance, post-Renaissance and modern.

P. 359: Why did early Christian thinkers reject Greek arguments for a round earth?

P. 361: How was astronomy recognized as useful during the reawakening of Europe after the Dark Ages?

Why was the information from Ptolemy's Almagest found inadequate during the Middle Ages?

P. 362: Did the Alfonsine Tables prove to be an adequate replacement for Ptolemy's Almagest?

How did Chaucer use astronomy in literature?

P. 363: What aspects of recovered Greek cosmology became unacceptable to the Holy Fathers?

What aspects fit well the Christian theology?

How did Thomas Aquinas avert a crises for the further development of astronomy?

How did the poet Dante incorporate astronomy into his epics?

P. 367: How did Dante (and other Christian thinkers) modify Greek cosmology?

How did Venice become the cultural capital of Europe in the late Middle Ages and the Renaissance?

P. 368: What events stimulated the development of navigation in Mediterranean waters?

P. 369:How did Greek and Alexandrian astronomy aid the navigators?

P. 370:What motivated Prince Henry of Portugal to navigate around the continent of Africa?

P. 370:How did Prince Henry advance navigation?

P. 371:Why did Portugese navigators reject Columbus proposal to sail west over a round earth to the reach the riches of the east ?

P. 372:How did the navigators' push to explore new frontiers advance astronomy?

Why was the Julian calendar becoming inadequate for the prediction of Easter?

P. 373:How does the current definition of the date of Easter depend in the lunar and solar calendars?

P. 375:What were some of the factors that allowed the Protestant Reformation to take hold in Germany?

P.376: How did the Roman Catholic Church resist the Reformation movement?

P.377: What problems did Copernicus have with the Ptolemaic system?

P. 378:How did the motions of Venus and Mercury inspire Copernicus to re-arrange the planetary system?

P. 380:How did Copernicus explain the long-known fact that Mars, Jupiter and Saturn can appear either in opposition or in conjunction with the sun, as distinct from Venus and Mercury?

P. 381:How did Copernicus explain that Venus can appear as an evening star for some months and as a morning star for other months and is totally invisible in between?

P. 382:How did Copernicus explain retrograde motion?

Why do planets appear brightest when these are in retrograde motion?

P. 383:In what ways was Copernicus heliocentric model unifying, simplifying and symmetric?

How did a change of perspective help Copernicus to simplify celestial motion?

P. 384:How did Copernicus overturn traditional order?

What are some of the parallels between Luther's Reformation movement and Copernicus' Revolution?

P. 385:How did late Renaissance and Baroque artists upset traditional Renaissance artistic thinking and style?

What is the reason for the seasons in the heliocentric system?

Why does the North star stay in a fixed position while other stars rotate around it through the course of one night?

P. 386: How do we know that the moon revolves around the earth?

What is the cause of the precession of the spring equinox?

P. 387: What were the new predictions from Copernicus' rearrangement?

P. 389: Why did Copernicus' contemporaries reject his predictions?

Think of another example of a prediction fulfilled posthumously.

P. 390: What were the deficiencies of Copernicus' simple heliocentric model?

P. 391: Why did Copernicus hesitate to publish his new ideas through the printing press?

Why did Martin Luther, who reformed religion, oppose the reformation of cosmology?

P. 393: What were the rational objections raised against the heliocentric system?

How did Copernicus answer these objections?

What rational objections did Copernicus fail to answer?

P. 394: What was Copernicus's answer for the absence of a parallactic shift in the position of the stars?

P. 395: Why did the Church allow the use of the Copernican system for re-computation of the calendar if they did not support its validity?

P. 396: According to the Gregorian calendar, will the year 3000 be taken as a leap year?

P. 397: Why was Copernicus' work not just a copy of Aristarchus' earlier heliocentric model?

How was Rheticus' work helpful to the further acceptance of the heliocentric system?

P. 398: How was Osiander's preface harmful to the further acceptance of the heliocentric system?

P. 399: What was the general reaction to Bruno's imaginative consequences of the Copernican system?

What part of Bruno's writings got him into deep trouble with the Church?

P. 400: How did Foucault get the idea for his famous pendulum experiment to prove the rotation of earth?

P. 401: Would Foucault's pendulum demonstrate earth's rotation if carried out in Equador?

P. 403: How did Bradley determine the direction "straight up" to point his telescope toward the zenith?

Besides proving the motion of the earth around the sun, what other important measurement did Bradley accomplish through starlight aberration?

P. 405: What was the first evidence that stars also move?

What property did Herschel discover about binary stars?

P. 406: What asymmetry did Herschel find that showed him that the sun also moves?

Chapter Questions

1. How did Greek cosmology influence Christian cosmology?
2. Point out the difference between the Copernican and Ptolemaic system for each of the following phenomena
   a. cycles of day and night
   b. nightly motion of the stars around Polaris
   c. seasons on earth
   d. movement of the heavenly bodies among the zodiac constellations
   e. lunar and solar eclipses
   f. precession of the equinoxes
   g. retrograde motion
3. How does the heliocentric theory account for (a) retrograde motion of the planets and (b) regular variations in the brightness of the stars?
4. Comment on whether Copernicus' heliocentric model was scientifically convincing.
5. What were Copernicus' arguments against the Ptolemaic system?
6. How did Copernicus improve the system of heavenly bodies in terms of: unity, symmetry, simplicity, elegance?
7. What is the difference between inner planets and outer planets in the Copernican system.
8. What were the deficiencies of the Copernican system? How did he try to resolve these?
9. Give one new prediction of the Copernican system that was successfully verified later?
10. What were the philosophic implications of the Copernican system that made it difficult for the scholastics and clerics to accept the heliocentric view?
11. Why did Copernicus hesitate to publish his system?
12. How did Renaissance artists reflect the centrality and dominance of earth (humanity) in the prevailing cosmology?
13. How did Baroque artists depart from the above views?
14. If the earth is moving (as we so readily accept today) why does an apple hit the ground just below where it sits on the tree?

15. List the first practical benefits to be realized from the new system.

16. The following are the flags for the Islamic countries (left) Pakistan and (right) Algeria. Are these pictures astronomically correct? If not, why not?

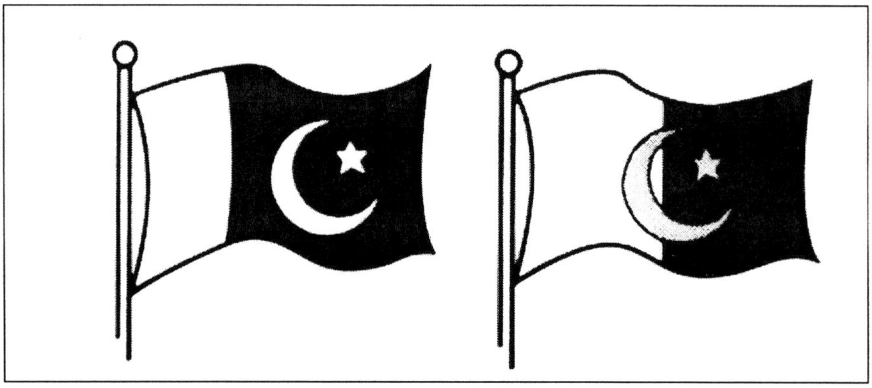

17. What is aberration of starlight?

18. How does the annual revolution of the earth around the sun help to measure the velocity of light?

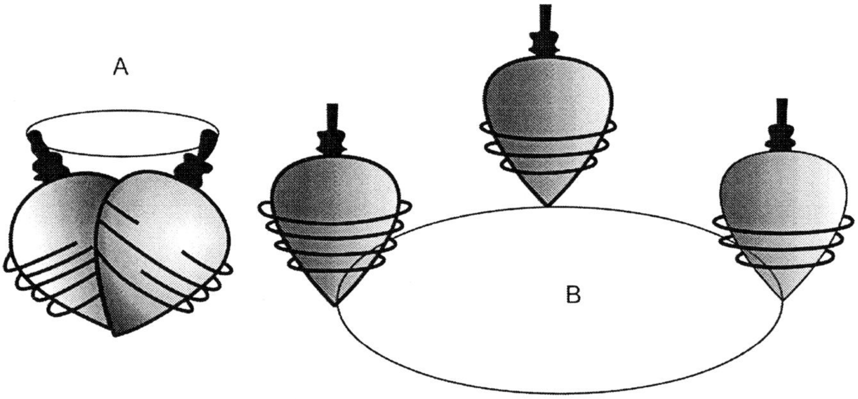

19. Which one of the above two motions represents precession of a moving top that resembles the precession of earth's axis of rotation.

20. Cite three separate parallactic effects that arise from observations by a moving observer.

21. An experimental airplane is designed to hover stationary above the earth at 1000 meters. In the meantime, the earth rotates on its axis and the plane descends to land when its destination appears below. Will such an airplane work for traveling long distances? Discuss your answer.

22. Cite an example of modern evidence that our earth rotates daily around its axis? Cite an example of modern evidence that earth revolves annually around the sun?

23. How does a swinging pendulum help prove the earth's rotation?

Math-Based Questions

1. The solar year (equinox to equinox) is 365.2422 days, not 365. 25 days. Convert the difference to minutes and seconds.

2. Show that adding a leap year every four years would still create a problem of overcorrecting by roughly one day every century.

3. Therefore the Gregorian scheme, which we use, skips the leap year on years that end in 100. For example, the year 1800 and 1900 were *not* leap years. Show that this will still leave a smaller discrepancy of about 1/4 day every 100 years.

4. The next order correction is to add in a leap year in a century year, if the century year is divisible by 400. This means that the year 2000 would still be a leap year, but 2100 would not. If this scheme is used, how long will it take for an error of one day to accumulate?

5. Draw two circles whose diameters are proportional to the sizes of the heliocentric orbits of earth and Venus. Using a protractor, measure the maximum angle that Venus can be away from the sun. Do the same for the planet Mercury.

6. Make a scaled drawing of the solar system from Copernicus' Table 4.1.

7. List the distances (in km) from the center of the earth to the following: Moon, Mercury, Venus, Sun, Mars, Jupiter, and Saturn, i.e. to the seven heavenly bodies.

8. Explain why the maximum angle between the sun and the planet Mercury is smaller than the angle between the sun and the planet Venus? Draw diagrams.

9. Calculate the orbital speed of the earth around the sun in meters/second. Radius of earth = 6400 km, Earth sun distance = 150 million km.

10. Calculate the angle of parallax to the star Sirius due to the earth's revolution around the sun.

11. Suppose the earth-sun distance equals one cm. How long a piece of paper would you need to represent the distance to the nearest star?

12. Look up the diameter of the earth's orbit in the Table p. 146. Calculate the circumference of the orbit. Now calculate the orbital speed of the earth in meters/second.

13. Repeat the above exercise for the rotation speed of a point on the equator of the earth in meter per second.

Exploration Topics

1. What were the arguments that were given against the heliocentric model of planetary motion when it was presented by Copernicus? Distinguish between the philosophical arguments and the scientific ones. Point out the flaws in the rational arguments against the heliocentric model.

2. Comment on the processes of the democratization (or instead, the specialization) of knowledge that we are witnessing in our present society. Make comparisons with parallel developments in other periods, for e.g. Egyptian, Greek, Medieval. Speculate as to what the future consequences are likely to be in the scientific or other arenas.

3. Discuss the developments in this chapter in terms of Einstein's statement: "Everything should be made as simple as possible, but not simpler."

4. Discuss how Greek ideas influenced Christian theology.

5. Modern astronomers are looking for planets around stars other than our sun. About how many planets have they found so far? zero, less than 10, about 50, about 100, thousands.

Sample Quiz

1. Islamic astronomers determined the more exact length of the year to be 365.2422 days instead of the 365.25 days which the Julian calendar used. If the Church kept using the Julian calendar from the Roman times till the year 1500, calculate how many years it would take for an error of one day to accumulate? Chose the closest answer.

   a. 100 years b. 1000 years

   c. 10 years d. 10,000 years

2. Which of the following statements is false?

   a. Venus can appear only in conjunction with the sun.

   b. Mercury can appear only in conjunction with the sun.

   c. Mars can appear only in conjunction with the sun.

   d. Jupiter can appear in conjunction or in opposition with the sun.

   e. Saturn can appear in conjunction or in opposition with the sun.

3. Cite an example of modern evidence that earth revolves annually around the sun?

   a. The cycle of seasons has a period of one year.

   b. The sun appears to move annually among the constellations of the zodiac.

   c. Some of the nearest stars appear to shift relative to far away stars by a fraction of an arc-second from winter to summer.

   d. The location of the sun on horizon changes in an annual cycle.

4. Calculate the orbital speed of the earth around the sun in meters/second. Chose the closest answer. Radius of earth = 6400 km, Earth sun distance = 150 million km. Chose the closest ans.

   a. 30,000 m/s b. 3000 m/s

   c. 300,000 m/s d. 300 m/s

5. Calculate the angle of parallax (half angle definition) to Mars from two places on earth that are located at the poles when Mars appears *brightest* to earth: (Mars-sun distance = 228 million km).

   a. 17 arc-sec b. 34 arc sec

   c. 8.5 arc sec

6. Suppose the earth-sun distance (8 light-minutes) equals one cm. How long a piece of paper would you need to represent the distance to the nearest star (5 light years away) ? Chose the closest ans.

   a. 30 m b. 300 m

   c. 3000 m d. 30,000 m

7. Identify the chart below that shows when Mars will be in retrograde motion

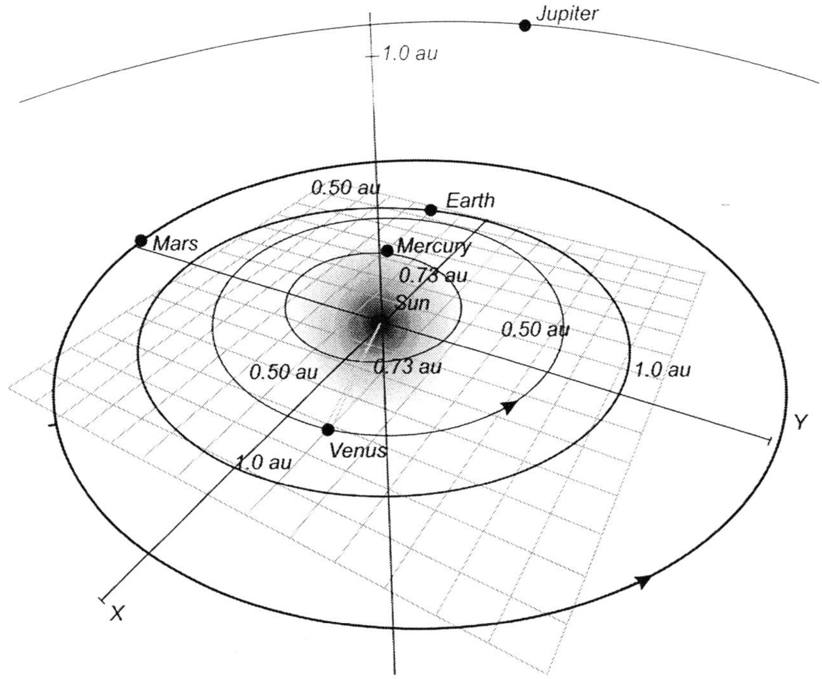

(a)

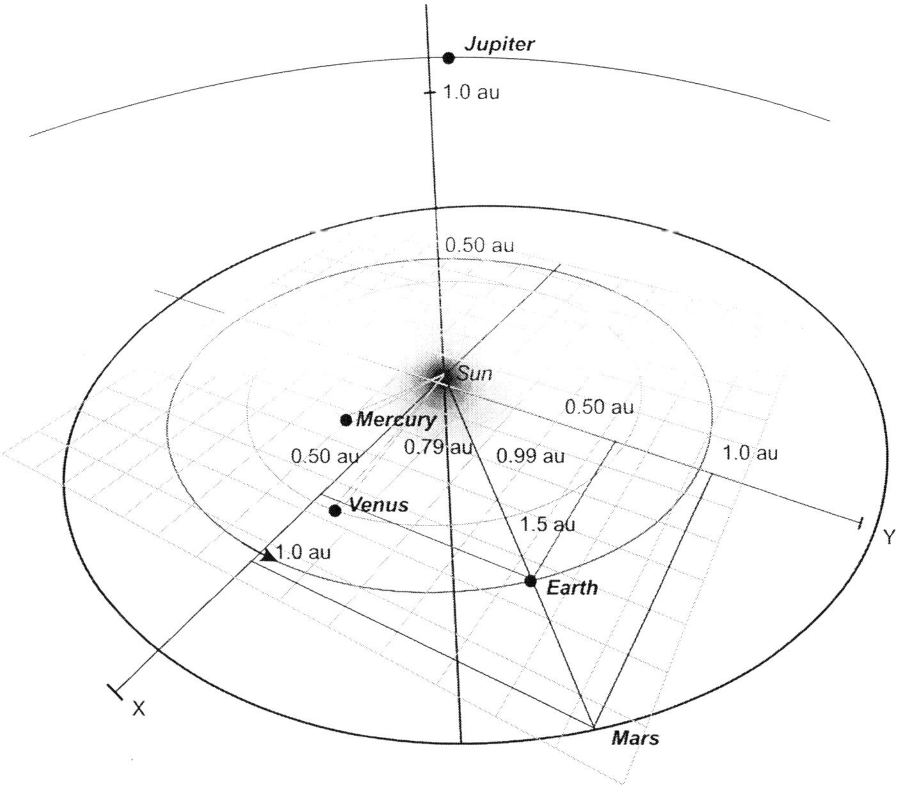
Jupiter
1.0 au
0.50 au
Sun
Mercury
0.50 au
0.79 au
0.99 au
0.50 au
1.0 au
Venus
1.5 au
Y
1.0 au
Earth
X
Mars

(b)

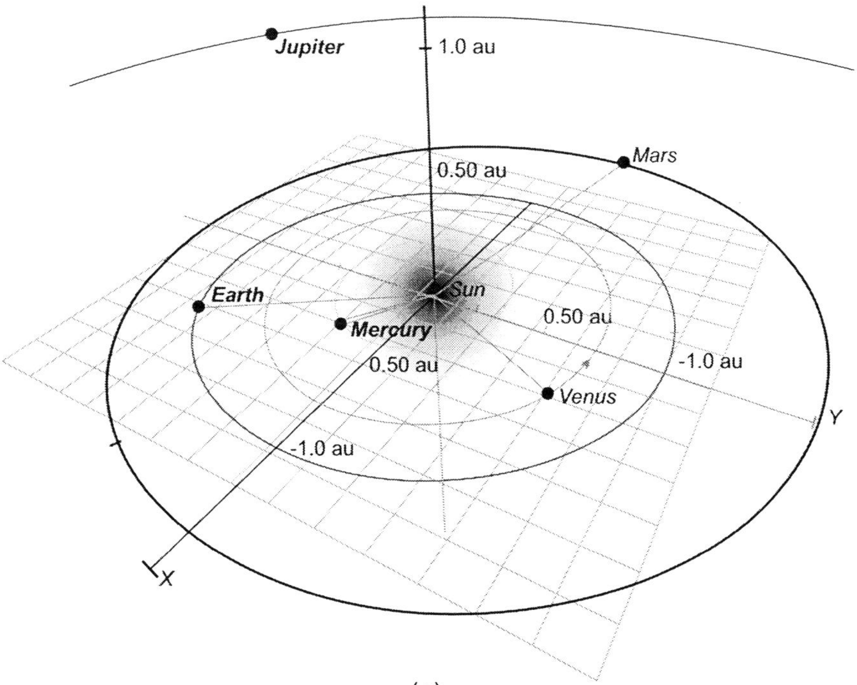
Jupiter
1.0 au
Mars
0.50 au
Earth
Sun
Mercury
0.50 au
0.50 au
-1.0 au
Venus
Y
-1.0 au
X

(c)

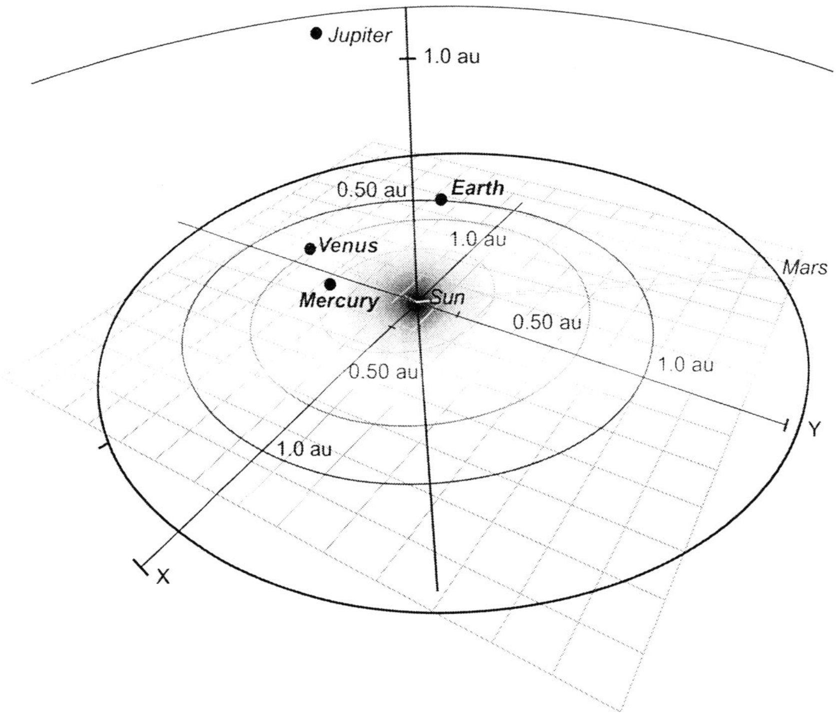
Jupiter
1.0 au
0.50 au
Earth
1.0 au
Venus
Mars
Mercury
Sun
0.50 au
0.50 au
1.0 au
Y
1.0 au
X

(d)

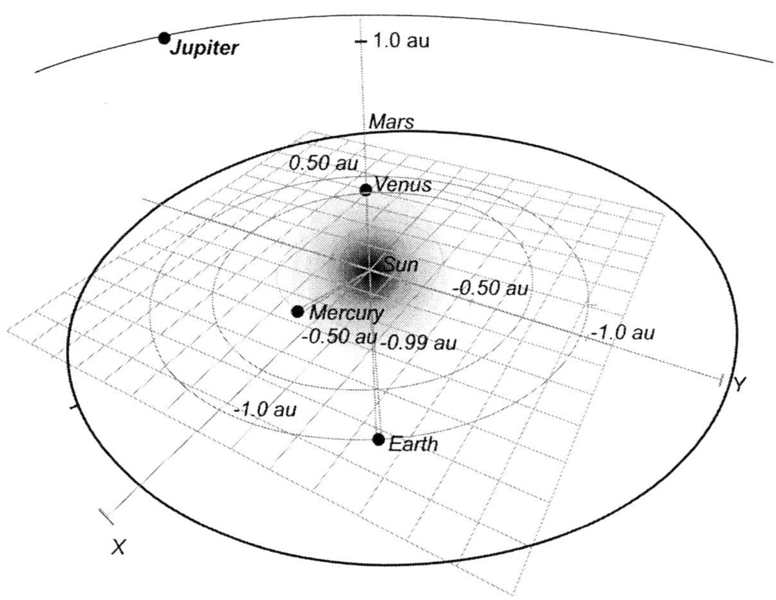
Jupiter
1.0 au
Mars
0.50 au
Venus
Sun
-0.50 au
Mercury
-0.50 au
-0.99 au
-1.0 au
Y
-1.0 au
Earth
X

(e)

Answers to Sample Quiz for Chapter 7

1. b
2. c
3. d
4. a
5. d
6. a
7. c
8. a

Homework Questions for Chapter 7 (with sample answers)

1. Point out the difference between the Copernican and Ptolemaic system for each of the following phenomena

   a. cycles of day and night

   b. nightly motion of the stars around Polaris

   c. seasons on earth

   d. movement of the heavenly bodies among the zodiac constellations

   e. lunar and solar eclipses

   f. precession of the equinoxes

   g. retrograde motion earth.

2. Comment on whether Copernicus' heliocentric model was scientifically convincing.

3. What were Copernicus' arguments against the Ptolemaic system?

4. What is the difference between inner planets and outer planets in the Copernican system.

5. What were the deficiencies of the Copernican system? How did he try to resolve these?

   There were a few deficiencies in Copernicus model. First, his original cosmology with only one circle for each planet was inaccurate. To get better quantitative results, he had to modify his model to include 34 circles, increasing its complexity. Next, was the problem of earth's motion. If the earth was moving, then people expected to be able to feel its motion, and expected this motion to cause a violent wind from the earth moving through the atmosphere. Copernicus resolution to the wind problem was to say that the air must be moving along with the earth. There were also similar objections from a belief that a rapid rotation would cause the earth to break apart. Copernicus pointed out, however, that the rotation of the earth was much less violent than the alternative daily rotation of the entire universe.

6. Give one new prediction of the Copernican system that was successfully verified later?

7. The solar year (equinox to equinox) is 365.2422 days, not 365. 25 days. Convert the difference to minutes and seconds. Show that adding a leap year every four years would still create a problem of overcorrecting by roughly one day every century.

8. Calculate the orbital speed of the earth around the sun in meters/second. Radius of earth = 6400 km, Earth sun distance = 150 million km.

To calculate the orbital speed of the earth around the sun we turn to our old friend $d = vt$. To determine earth's velocity, therefore, we need only know a distance it travels (say, once around the orbit?) and the amount of time that takes (one year!), and we are done, We make the reasonable assumption that Earth's elliptical orbit is essentially circular in order to declare the distance traveled in one year to be the circumference of a circle with radius Rs,E. With this assumption we find

$$v = \frac{d}{t} = \frac{C_{S,E}}{t} = \frac{2\pi R_{S,E}}{t}$$

$$= \frac{2\pi \times 150 \times 10^6 \text{km}}{1 \text{ year}}$$

$$= \frac{2\pi \times 150 \times 10^9 \text{m}}{1\text{year}\left(\frac{365.25 \text{ days}}{1 \text{ year}}\right)\cdot\left(\frac{24 \text{ hours}}{1 \text{ day}}\right)\cdot\left(\frac{60 \text{ minute}}{1 \text{ hour}}\right)\cdot\left(\frac{60\text{s}}{1 \text{ hour}}\right)}$$

$$= \frac{9.4 \times 10^{11} \text{m}}{3.16 \times 10^7 \text{s}}$$

$$= 29900 \text{ m/s}$$

9. Cite an example of modern evidence that our earth rotates daily around its axis? Cite an example of modern evidence that earth revolves annually around the sun?

10. From the modern value of the earth's radius, determine the rotation speed of a point on the equator of the earth in meters per second.

CHAPTER 8

Reading Questions

P. 410: What aspect about the total solar eclipse stimulated Brahe's deep interest in astronomy?

P. 412: How did Brahe decide to spend his inheritance from his dead uncle?

How did Brahe know that the bright object is Cassiopea must be a new star?

What observations did Brahe make about the nova? How long did the new star remain bright enough to be visible?

What difficulties did the new star present to Aristotle's widely accepted ideas?

P. 414: How did Brahe improve accuracy from Hipparchus' 10 arc-minutes to 1 arc-minute?

P. 416: What are the parallels between Brahe's approach to precision measurements and the Renaissance artists approach to precise description of nature?

P. 417: Why did Brahe reject Copernicus' heliocentric model?

P. 418: How did Brahe determine that the new comet belonged to the heavens and not to the atmosphere?

How far was the comet relative to the moon?

P. 422: What patterns did Kepler thought he discovered in the planetary system?

P. 426: How did geometrical patterns influence the work of Renaissance artists like Michelangelo.

Why was Kepler dissatisfied with his geometric planetary model based on perfect solids?

P. 427:Why did Brahe give little credence to Kepler's model of perfect solids?

Why was Brahe so grudging in sharing his data with Kepler after he invited him to collaborate?

P. 429:Why was Kepler dissatisfied with the fit of Mars orbit composed from multiple circles to the data of Brahe?

What was the underlying reason for the failure of 2000 years of astronomy to model planetary orbits with combinations of circles?

P. 430:What was the extent and significance of Kepler's success in using elliptical orbits for the planets?

What is Kepler's first law of planetary motion?

P. 431:Besides using ellipses, how did Kepler's new model depart from Copernicus's sun-centric system?

Point out the parallels between Kepler's departure from the classical/ Renaissance approach and the innovations of the Baroque artists.

P. 434:Why was Kepler still unsatisfied with the success of the elliptical construction for the planetary system?

P. 435:Why do some total solar eclipses leave an annular ring of light around the sun?

P. 436:How does the speed of a planet vary as it orbits the sun?

P. 437:How did Kepler unify the motion and orbits of all the planets?

How did Kepler fulfill Pythagoras's ambition?

P. 440:From his laws of planetary motion, what did Kepler conclude about the force between the sun and the planets?

P. 442:How long did Kepler's nova stay bright enough to be visible?

P. 446:What types of stellar objects can the nebulae be classified into?

Why do all planets revolve around the sun in the one direction, and also rotate around their axes in that same direction? Are there exceptions?

P. 447:Where did the hydrogen and helium in the interstellar gas clouds come from originally?

P. 448:What kind of events can start the rotation in a part of the interstellar gas clouds?

Why does a spinning spherical cloud change into a disk shape?

How do planetary masses form as a star pulls together?

P. 449:What conditions are necessary for the contracting core to turn into a star?

P. 450:How does the earth continue to grow slightly in mass?

What are the differences between the Oort cloud and the Kuiper belt?

P. 451:How will the orbit of a comet from the Oort cloud be different from the orbit of a comet originating from the Kuiper belt?

What is the cause of a comet's tail? Does the tail point in any special direction?

P. 452:What caused the extinction of the dinosaurs and other species 65 million years ago? What is the evidence for the proposed cause?

Which elementary particles are generated in a nuclear fusion reaction that powers the sun?

Why does the sun not collapse due to gravity?

P. 453:Why does the sun not blow up from the many nuclear fusion explosions?

P. 454:What is a red giant star? How does it form?

Where do the light elements (up to iron)of the Periodic Table come from?

P. 455:Why is fusing iron into the heavier element harmful to the life of the star?

Why does a neutron star not collapse any further?

P. 456:Compare the magnetic field strength of the neutron star with the earth-magnet.

In what form does a star lose most of its energy when it explodes into a supernova?

P. 457:Where do the elements (heavier than iron) come from?

How does nuclear physics help to understand astrophysical phenomena?

How do humans owe our existence to the life and death of stars?

P. 458:Compare Zwicky's quest to Brahe's quest.

P. 459:Why did neutrino detectors fail to signal the occurrence of SN1987a?

P. 460:How do we know that there must be a neutron star at the center of SN1987a remnant if it is not a pulsar?

P. 461:Which supernova remnant clearly shows a pulsar?

Math Based Topics

Planetary Properties

Table 1:

	Earth	Moon	Sun
Mass kg	6×10^{24}	7.4×10^{22}	2×10^{30}
Mean Radius m	6.4×10^{6}	1.74×10^{6}	7×10^{8}
Mean density kg/m^3	5.5×10^{3}	3.3×10^{3}	1.41×10^{3}
Orbital Period days	365.242	27.3	
Mean distance from Sun m	1.5×10^{11}		
Mean distance from Earth m		3.85×10^{8}	

Planets	**Semi-Major Axis (10^6 km)**	**Period (years)**	**Mass/Earth's mass**
Mercury	57.9	0.24	0.055
Venus	108.2	0.615	0.815
Earth	149.6	1.0	1
Mars	227.9	1.88	0.107
Jupiter	778.3	11.86	317.9
Saturn	1427.0	29.5	95.2
Uranus	2871.0	84.0	14.6
Neptune	4497.0	164.8	17.2
Pluto	5913.5	248.5	0.02

Kepler's Law 1:

All planetary orbits are ellipses with the sun at one focus.

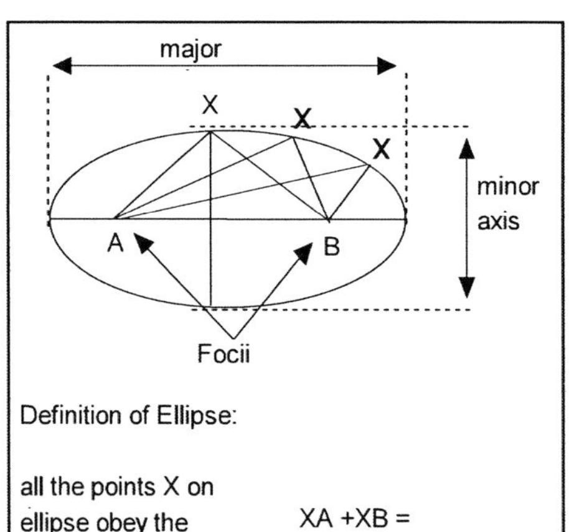

Fig. 8.1: Definition of an ellipse.

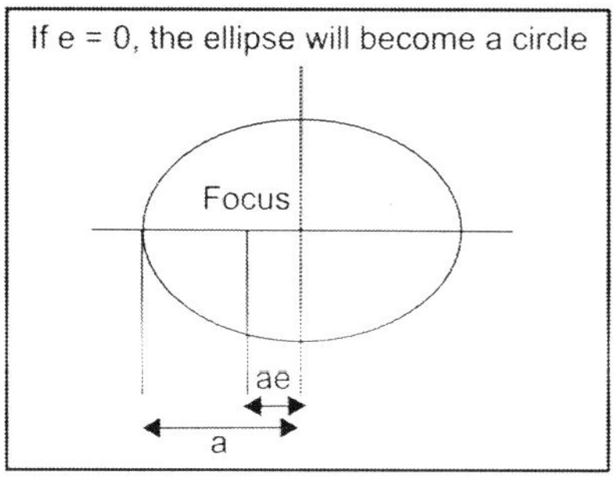

Fig. 8.2 The definition of the eccentricity (e) of an ellipse. The semi-major axis is a. The product (ae) is the distance of the focus from the center.

Table 2: The planetary orbits are nearly circles, as shown by the slight eccentricity of the ellipses, except for Mercury.

Planet	Eccentricity e
Mercury	0.206
Venus	0.007
Earth	0.017
Mars	0.0093
Jupiter	0.048
Saturn	0.056

Kepler's First Law of Motion of Planetary Motion:

All planets describe elliptical orbits with the sun as one of the foci.

Kepler's 2nd Law

Planets sweep out equal areas in equal times over their orbits.

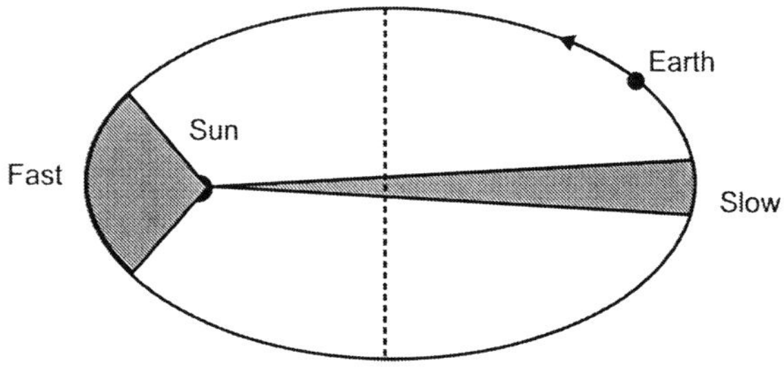

Fig. 8.3 Kepler's equal area law. The sun is at the focus of the ellipse. When a planet is closer to the sun, it moves faster than when it is distant from the sun. As a result, the line connecting the planet to the sun will trace out the same area in the same time interval.

Kepler's 3rd Law (The Harmonic Law)

The square of the orbital period T is proportional to the cube of the semi-major axis R of the orbital ellipse.

R^3/T^2 is a constant

Table 3 Application of Kepler's 3rd Law to the six planets.

Planet	Relative (Distance from the sun)	Period in Years	Square of Period	Cube of Distance
Mercury	0.387	0.241	0.058	0.058
Venus	0.723	0.615	0.378	0.378
Earth	1	1	1	1
Mars	1.524	1.881	3.538	3.54
Jupiter	5.203	11.862	140.707	140.851
Saturn	9.539	29.458	867.774	867.977

Example

Two new planets, Uranus and Neptune were discovered after Newton. The semi-major axis of Uranus' orbit (relative to the earth's orbit) is: 19.19. Using Kepler's 3rd Law, calculate the periods (in years) of Uranus.

$$R^3 = 19.9^3 = 7066.8$$
$$T^2 = 7066.88$$
$$T = \sqrt{7066.8} = 84 \text{ years}$$

Example

Using Kepler's law and the 75 year period of Halley's comet, calculate the average orbital radius.

According to Kepler's law, using R = 1 and T = 1, for earth,

$$\frac{R^3}{T^2} - 1 \qquad R^3 - T^3$$
$$T^2 = 75^2 = 5625 = R^3$$
$$R = ^3\sqrt{5625} = 17.7$$

Nearly twice as larger as R for Saturn = 9.55 (Table 1).

38. Did the element uranium form inside a star or during a supernova?

39. Why is it important to know the mass of star?

40. Is our sun a lonely star or is it part of a cluster?

41. Pick a number (1 to 4) below the stage of stellar evolution as the stages occur during the birth of a star.

    Protostar, Fusion begins, Interstellar cloud, Central clump forms with an accretion disk

Math Based Questions

1. Hold out your thumb at arms length against a background of distant objects. Examine it with one eye closed, then the other eye closed. Describe the position of your thumb relative the distant background. Estimate the distance between your eyes and the distance from your face to your thumb. Calculate the angle of parallax.

2. Now move your thumb to a position about 10 cm from your face. Examine it with one eye closed, then the other eye closed. Describe the position of your thumb relative the distant background. Calculate the new angle of parallax. Similarly the parallax effect can be used to find the distance to the sun, the moon, comets, and stars.

3. In one version of Kepler's law for the solar system, R^3/T^2 is a constant, equal to unity.

   What are the units for R?

   What are the units for T?

   What happens to the value of Kepler's constant if the law is applied to the orbit of the moon? Calculate the value if it is different.

4. Two new planets, Uranus and Neptune were discovered after Newton. The semi-major axis of Neptune's orbit (relative to the earth's orbit) is: 30.06. Using Kepler's 3rd Law, calculate the periods (in years) of Neptune.

5. Construct (a) a circle using a piece of string, a pencil and one thumb tack for the fixed center (b) an ellipse using a piece of string, a pencil and two thumb tacks for the two foci. Show the foci, major axis, minor axis. What is the value of the eccentricity of the ellipse that you have chosen to construct. Now construct another ellipse with a greater eccentricity.

6. Draw a solid dot whose diameter is such that it subtends an angle of one arc second at distance of 1 km.

7. Using the estimate of the distance between the earth and the moon, calculate the angle of parallax to the moon from two places on earth separated by 200 km. Tycho claimed that the comet was 6 times farther from the earth than the moon. What was the angle of parallax that he would have measured for the comet from the same two places on earth, i.e., separated by 200 km. If his *best* accuracy was 0.5 minutes of arc, do you believe his claim?

8. Calculate the annular stellar parallax using Ptolemy's estimate of 20,000 earth radii for the distance to the stellar sphere and 1210 earth radii for the earth to sun distance. The annual displacement should be nearly 7 degrees! The constellations would noticeably change shape over the course of a year.

9. What is the distance between two diametrically opposite points on the earth's heliocentric orbit? From these two points, calculate the angle of parallax to (a) the sun (b) the nearest star, assuming that it is 200,000 times as far away from the earth as the sun. Estimate how far you would have to place a dime from you to get the same parallax angle as the one to the nearest star. Compare the distance you obtain to a geographical size (e.g. the distance from one U.S state to another).

10. The earth is closest to the sun in January and farthest from the sun in July. In which month is the earth traveling fastest in its orbit?

11. Suppose that a comet located half way between Mars and Jupiter shows a tail that subtends an angle of 22 degrees to an earth observer. Draw a heliocentric diagram showing the positions of the sun, the earth, Mars, the comet and Jupiter. Calculate the length of the comet's tail in astronomical units, using the information given in Table 1, pg 146

12. Compare the width of the 1577 comet with the diameter of the moon. Remember to take into account the distance of the comet from the earth.

13. The planetoid Ceres goes around the Sun once every 2.6 years. What is the radius of its orbit measured in terms of the radii of the earth's orbit around the sun? Use Kepler's third law, R^3/T^2 is a constant.

14. Which of the following takes the longest time to complete an orbit around the Sun?

    a. Jupiter b. Saturn

    c. Mars d. Earth

15. Two new planets, Uranus and Neptune were discovered after Newton. The semi-major axis of Neptune's orbit (relative to the earth's orbit) is: 30.06. Using Kepler's 3rd Law, calculate the periods (in years) of Neptune.

16. Calculate the annular stellar parallax using Ptolemy's estimate of 20,000 earth radii for the distance to the stellar sphere and 1210 earth radii for the earth to sun distance. Would that be observable by Hipparchus? By Tycho?

17. From the semi-major axis (radius) of the orbit of Io, and the period of each of the other four satellites of Jupiter, calculate the radius of the orbit of Europa and Ganymede in the units of 1000 km. Be careful with units.

Satellites of Jupiter	**Radius of Orbit (10^3 km)**	**Period**
Io	422	1.77 earth days
Europa	3.55 days	
Ganymede	7.15 days	

Exploration Topics

1. Discuss the developments in this chapter in terms of Whitehead's quote at the beginning of Chapter 5.

2. Discuss the developments in this chapter in terms of Einstein's statement:

   "Everything should be made as simple as possible, but not simpler."

3. Discuss the developments in this chapter in terms of quote at the beginning of the chapter.

   "Just as the ears were made for sound and the eyes for color, so the mind is made for quantity and precision. It is lost in darkness when it leaves the realm of quantitative thought."

4. About how many supernova explosions may have taken place within our MW galaxy since its birth?

5. What connections do the birth, life and death of stars reveal between the micro-world (of elements, atoms and nuclear physics) and the macro-world of stars and planets?

6. All stars do not eventually become supernovae with neutron stars. Discuss the different modes in which a star's life can evolve, and what properties of the star determine the course of its evolution.

Sample Quiz

1. Which of the following is true?

    a. As a planet makes its orbit around the sun, the distance of a planet from the sun is constant.

    b. The speed of a planet is constant over its orbit around the sun.

    c. The area traced by a spoke connecting a planet to the sun is constant for a constant time interval.

    d. All of the above.

    e. None of the above.

2. In one version of Kepler's law for the solar system, R^3/T^2 is a constant, equal to unity. What are the units for R?

    a. m b. km

    c. Astronomic Unit (AU) d. Light-Years

3. Halley's comet returns every 76 years. What is the semi-major axis of its orbit in terms of the radius of the earth's orbit around the sun

    a. 662.6 b. 17.9

    c. 76 d. none of the above

4. From the semi-major axis (radius) of the orbit of Io, and the period of the other four satellites, calculate the radius of the orbit of Europa in the units of 1000 km.

    a. 422 b. 671

    c. 1070 d. 1885

Satellites of Jupiter	Radius of Orbit (10^3 km)	Period
Io	422	42.5 hours
Europa		3.55 days
Ganymede		7.15 days

5. What is the source of the heat and light energy for our sun?

    a. chemical energy

    b. mechanical energy

    c. nuclear energy

    d. all of the above

6. Why is a supernova so much brighter than a star?

   a. Because it is nearer to the earth.

   b. Because it is the explosion of the star.

   c. Because it arises from the collision of two stars

   d. None of the above.

7. Most comets spend most of their time

   a. near the earth

   b. in the earth's atmosphere

   c. in the outer reaches of the solar system

   d. near Jupiter

8. Which of the following will not accompany a supernova?

   a. emission of trillions of neutrinos

   b. formation of a neutron star

   c. formation of elements heavier than iron

   d. birth of a galaxy

9. Why does a helium nucleus have a mass of one percent less than the combined mass of two protons and two neutrons?

   a. Because of the missing electrons.

   b. Some of the mass of the constituents is converted to energy during the formation of the helium nucleus.

   c. It is not possible to determine small nuclear masses to one percent accurary.

   d. None of the above.

10. Under which of the following conditions is a star is likely to become a supernova?

    a. It is more than 10 times massive as the sun.

    b. It is in the milky way.

    c. It uses up all its hydrogen.

    d. It moves very fast.

11. Which of the following will accompany a supernova?

    a. Emission of more than trillions of neutrinos.

    b. Formation of elements heavier than iron.

    c. Birth of a galaxy.

d. Formation of planets.

e. (a) and (b).

12. At what stage in the life of our sun will the planets be destroyed?

    a. When the core starts fusion reaction in lead.

    b. When all the helium at the core is exhausted.

    c. When all the hydrogen at the core is fused into helium.

    d. With the formation of a neutron star.

    e. With the release of a trillion trillion neutrinos.

Answers to Sample Quiz for Chapter 8

1. c
2. c
3. b
4. b
5. c
6. b
7. c
8. d
9. b
10. a
11. e
12. c

Homework Questions for Chapter 8 (with sample answers)

1. How did Tycho distinguish that the nova he saw was a star and not a comet? How did Tycho's level of accuracy compare to Hipparchus accuracy? Could Tycho measure the annual parallax of the stars?

2. How do we know that comets belong to the heavens and not the atmosphere?

3. How did Tycho's level of accuracy compare to Hipparchus accuracy?

4. How did Kepler's laws make the description of the universe more elegant?

5. From the semi-major axis (radius) of the orbit of Io, and the period of each of the other four satellites of Jupiter, calculate the radius of the orbit of Europa and Ganymede in the units of 1000 km. Be careful with units. Show your work.

Satellites of Jupiter	**Radius of Orbit (10^3 km)**	**Period**
Io	422	1.77 earth days
Europa		3.55 days
Ganymede		7.15 days

Solution

Satellites of Jupiter	Radius Orbit (10^3km)	Period (earth days)
Io	422	1.77
Europa	671	3.55
Ganymede	1070	7.15

6. What ancient observations show that the all planetary orbits lie approximately in one plane? What observations about planetary motion support the idea that the solar system formed from a rotating cloud of interstellar gas?

7. Besides using ellipses, how did Kepler's new model depart from Copernicus's sun-centric system?

   Why was Kepler still unsatisfied with the success of the elliptical construction for the planetary system?

8. Why do all planets revolve around the sun in the one direction, and also rotate around their axes in that same direction?

9. Why does the sun not collapse due to gravity? Why does the sun not blow up from the many nuclear fusion explosions?

10. Which of the following play a role in the formation of a star?

    interstellar clouds,

    fission of the uranium nucleus,

    fusion of hydrogen nuclei into helium nucleus,

    unity of space and time,

    equivalence of mass and energy,

    gravitation,

    rotation

11. Where did the elements from the periodic table come from?

Solution

Stars and supernovae. The elements H and He formed soon after the Big Bang.

12. If 60 billion neutrinos rain down on every square cm of the surface of the earth from the sun' fusion reactions, determine how many neutrinos must leave the sun per second. You will need the earth-sun distance and the surface area of the earth.

CHAPTER 9

Reading Questions

P. 464: What kind of lens pair did Hans Lippershey first use to obtain a telescopic image of the church steeple?

P. 465: Why was Galileo interested in making practical devices?

What kind of lens pair did Galileo first use to obtain a telescopic image ?

How did his image differ in character from Lippershey's image?

P. 466: Why was Galileo able to make high quality telescopes without great distortion?

P. 467: How did Galileo deduce that there must be mountains on the moon just as on earth?

How did he treat this discovery?

Why is it possible to still make out the dark part of a crescent moon?

P. 468: Does the earth shine like a planet to a viewer from Mars?

What was the impact of Galileo's astronomical discoveries on artists of the time?

P. 469: What is a diamond ring eclipse?

What was troubling about Galileo's discovery of sun-spots to the intellectuals of his time?

P. 470: How did Galileo deduce that the sun rotates? How did he treat this discovery?

P. 472: Which conjectures of Copernicus was Galileo able to verify with his telescopic observations?

P. 473:How did the appearance of Venus confirm Copernicus heliocentric model and refute the earth-centric models?

P. 476:How did Galileo know that the new heavenly bodies he observed around Jupiter through the telecope were not new stars?

How did he treat this discovery?

P. 477:How did the moons of Jupiter help Galileo argue for the Copernican heliocentric system?

P. 478:How did individual stars appear through Galileo's telescope?

What did the telescope reveal about stars?

P. 479:What did Galileo discover about the Milky Way?

P. 481:Why did the Starry Messenger pamphlet become so popular?

P. 482:Why did the Church have mixed reactions to Galileo's discoveries?

What objections did he get from his academic colleagues?

What were the reactions of the contemporary intelligentsia?

P. 483:How did Galileo respond to the criticisms?

P. 485:How did Kepler advance Galileo's discovery about the moons of Jupiter?

P. 486:What did Galileo discover about Saturn? How did he treat this discovery?

Why did Galileo often report his discoveries in anagrams?

P. 487:Why did Galileo move from Padua to Florence?

What impact did this move have on his career?

P. 488:How did Galileo use Aquinas's approach to contest the Church's objections?

P. 489:Why did the Church discourage open interpretations of the scriptures?

P. 490:How did Galileo try to convince the new Pope about the Copernican view and answer the standard arguments ?

P. 491:Where did Galileo go wrong about his tidal explanations?

P. 492:What arguments did Galileo present to support his idea that heaven and earth are not two separate realms?

How did Galileo argue for the earth's motion even though all the observations indicated a stationary earth?

P. 493:How did Galileo try to satisfy the Pope's demand for a definitive conclusion supporting the Church's position of a stationary earth at the center?

P. 494:What were the various reactions to the publication of the Dialogs?

P. 495:How important was Galileo's tragic challenge to the Church for the eventual progress of science?

How did Galileo continue to push scientific advances while in exile?

P. 497:How did Galileo influence the next generation of scientists in the progress of science?

P. 498:How did Huygens advance the telescope? What new discoveries did he make?

What new discoveries did Cassini make about Mars and Jupiter?

P. 499:Why did Cassini choose Mars as the nearest planet suitable to make a parallax measurement rather than Venus which is actually closer to earth than Mars?

P. 501:Why did Galileo, and later Cassini, compile accurate tables for the eclipse times of Jupiter's moons?

P. 502:How did the measurement of the earth-sun distance play a role in determining the speed of light?

P. 503:Why does the Milky Way appear as a ribbon of light around the sky?

P. 505:Based only on lenses, what problems did refracting telescopes present to further advances in magnification?

What was the diameter of Newton's light collecting mirror?

How did Herschel increase the magnification power of the reflecting telescope?

What was the mirror diameter of Herschel's best reflecting telescope?

P. 506:What was Herschel's first major discovery through his powerful reflecting telescope?

What are island universes?

P. 507:How did Herschel open up the universe beyond the solar system and the Milky Way galaxy?

How did Herschel confirm Kant's hypothesis about the Milky Way?

P. 508:What was the diameter of the Leviathan telescope mirror?

Why was there doubt that the spiral nebulae are galaxies of stars like our Milky Way?

P. 509:Where did Kapteyn and Herschel put the location of the sun in their maps of the stars of the Milky Way?

P. 510:What was the mirrror diameter of Hale's telescope atop Mt. Wilson?

What are the major differences between open and closed star clusters which led Shapley to determine the full size of the Milky Way?

P. 511:How did Shapley determine distances to the globular clusters?

What asymmetry did Shapley find about the distribution of the globular clusters?

How did Shapley displace the sun from the center of the Milky Way?

What was the general reaction to his find?

P. 513:What evidence was showing up to suggest that the spiral nebulae must have stars?

During the Great Debate what evidence did Curtis offer to support that the spiral nebulae must be outside the Milky Way?

How were both Curtis and Shapley guilty of a bit of Ptolemaic thinking on the new galactic scale?

P. 514:How did Hubble determine the distance to the Andromeda galaxy?

P. 515:What was the main reason for Kapteyn's gross underestimate of the full size of the Milky Way galaxy of stars?

How did Trumpler determine the existence of galactic fog?

P. 516:Why was Shapley's method with globular clusters a better method to determine the size of our galaxy over Kapteyn's method?

Compare the time it would take light to cross our galaxy to the time of human existence on earth.

P. 517:Why do stars farther than the sun from the center of the MW galaxy move slower that stars closer to the center?

What is the period of revolution of the sun around the galactic center?

What is the distance of the sun to the galactic center (modern value)?

Math Topics

Lenses, Spectacles and Telescopes

How does a lens improve the failing eye? We use the techniques of ray tracing to show the principles underlying the use of lenses as optical devices, such as the magnifying glass, spectacles and the telescope. These principles were developed by Kepler after he learned about the telescope. In each case, we construct *an image* formed by the intersection of light rays as they travel from the object to the eye.

We have already discussed the phenomenon of refraction - the bending of light as it travels from one medium to another. A ray of light bends towards the normal when it crosses a straight boundary between an optically rare medium and a dense medium. Now we consider what happens when a ray crosses a curved boundary. If the ray is parallel to *the axis of lens* (Fig. 9.1), it enters the curved glass region and bends toward the axis of the lens. As we discussed earlier, it bends toward the normal by refraction. But now, in the case of the curved boundary, the normal is perpendicular to the tangent to the curve. When a parallel ray leaves the curved glass region, it bends *away from the normal*. But, because of the opposite curvature, the result is that the ray once again bends toward the axis of the lens. Therefore, if a convex lens is formed by joining two curved pieces of glass, any ray of light traveling parallel to the lens axis will bend toward the axis (Fig. 9.2). The net result of a convex lens is to *converge* parallel rays of light to a point of *focus*. A convex lens can thus be used to focus the light and heat rays of the sun to light a candle. For the same reason, any ray of light arriving at the convex lens from the focal point will emerge parallel to the axis of the lens.

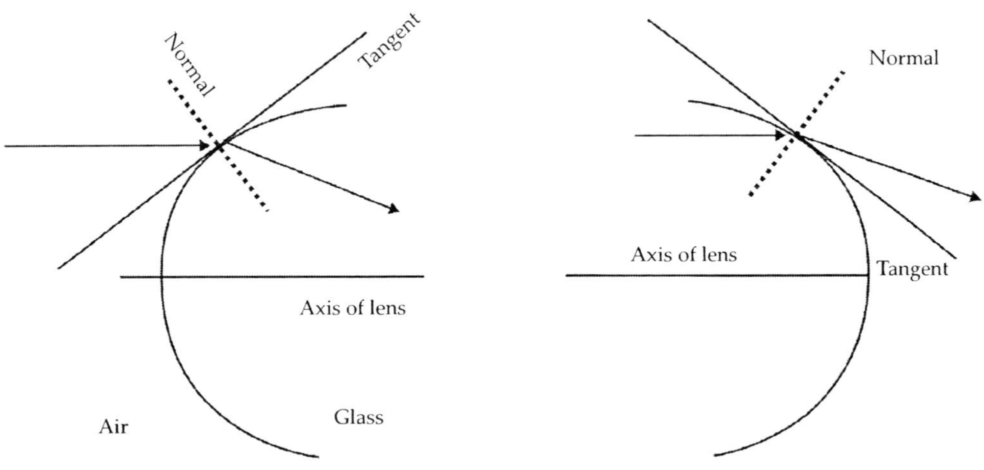

Fig. 9.1 Refraction of light through curved surfaces. Ray diagrams for the two faces of a convex (converging) lens.

Fig. 9.2 Ray diagrams for a convex (converging) lens. On the left, a parallel bundle of rays passing through a lens come to convergence at the focal point F. On the right, an object located at the *focal point* F emits light that emerges from the lens as a parallel beam. The convergence of parallel rays is one of the major principles used to construct ray diagrams that explain the operation of optical devices.

Fig. 9.3 Ray diagrams for a concave (diverging) lens. On the left, rays entering parallel emerge divergent, as if they come from F. On the right, rays heading for the focus, F, emerge parallel to the axis. The divergence of parallel rays is another major principle used to construct ray diagrams that help understand the operation of optical devices.

If two curved boundaries are joined together to form a *concave* lens, a similar ray tracing exercise will show that rays of light parallel to the axis will diverge away from the axis (Fig. 9.3).

The key organ for the operation of the eye is also a lens. For one type of abnormality, the rays come to a point in front of the retina - the light-sensitive receptor where all the optic nerves from the eye connect to the brain. This is called the *nearsighted* eye - a common problem with the adolescents. In another type of eye defect, the rays converge behind the retina. This is referred to as the *farsighted* eye - a common problem with the aging eye. Both cases cause blurred vision. Lenses change the path of the rays, so they come to focus exactly at the retina. The extra glass lens outside assists the eye's internal lens.

Ray tracing principles can be used to understand how a convex lens helps to correct the vision of an aging eye which has difficulty focusing on near objects - for reading. As mentioned, the incorrect (farsighted) eye focuses parallel light beyond the retina. The eye can accommodate by using its muscles to increase the focusing power (curvature) of its lens. As a result it can adjust to focus far away objects, as long as they do not get closer than the *near-point*. But when the object gets closer than the near-point, the eye can not strain enough to compensate, resulting in blurred vision for reading. (Fig.9.4). By adding convergence with a convex lens, distant objects can be focused easier on to the retina (Fig.9.4).

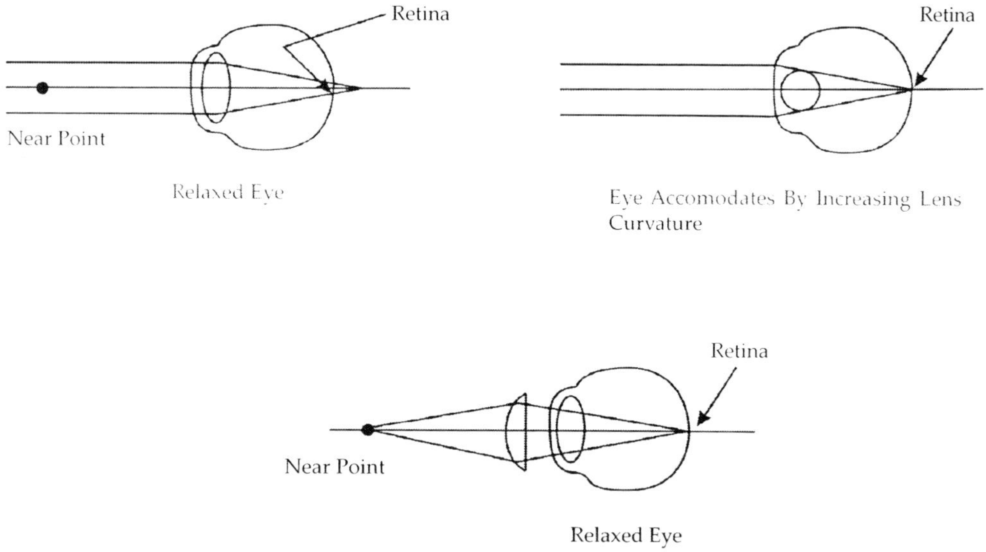

Fig. 9.4 Correction of the farsighted eye. A farsighted eye will focus parallel light from distant object to a point beyond the retina. The eye can still focus by accommodation as long as the object is not closer than the near-point. For nearer objects, a converging lens, increases the focusing power of the natural lens to bring the rays to a focus at the retina.

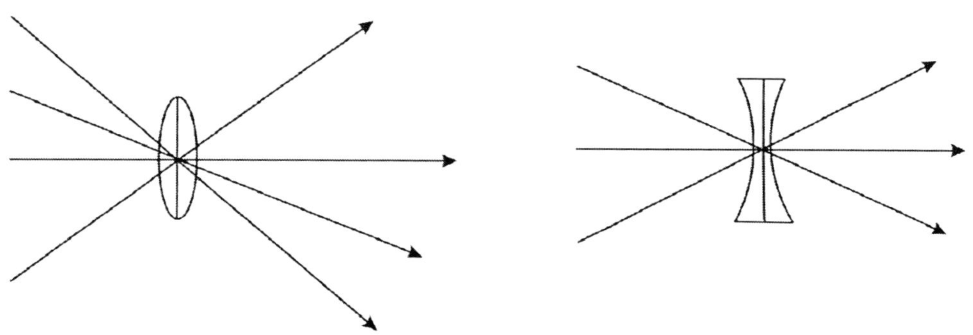

Fig. 9.5 An important ray tracing principle for convex and concave lenses. A ray of light that passes through the center of the lens travels un-diverted in a straight line.

A similar analysis applies to the nearsighted eye, which focuses the rays in front of the retina. The natural converging power of the eye is too strong. It can accommodate to focus near objects, as long as the distance does not exceed the *far-point*. Beyond the far-point, the image is always blurred. In this case diverging lens helps to fan out the rays to focus at the retina.

Another important principle for ray-tracing optics is that any ray that travels through the center of a lens - convex or concave - will pass through without bending. This is because the bending in one direction upon entry is compensated for by bending in the opposite direction on exit (Fig. 9.5). We use this principle to show how a convex lens forms a miniature *real image* of an object located beyond the focal point (Fig. 9.6). The image can illuminate a screen or a piece of white paper placed at the location of the formed image. It is this property which makes the convex lens into a burning glass; the lens focuses the light rays from the sun to form a real image of the sun.

If the object is located farther from the lens than the focal point, the convex lens serves as a magnifying glass bringing the image nearer and magnifying it. A concave lens also forms a *virtual image* (Fig. 9.7). When the light rays arrive at the eye, the eye-brain system comprehends them as arriving from a non-existing image (Fig. 9.7).

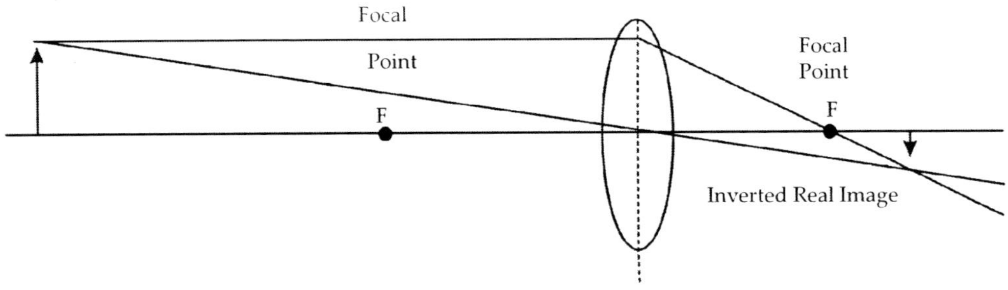

Fig. 9.6: A convex lens forms a real and inverted image of an object located beyond the focal point.

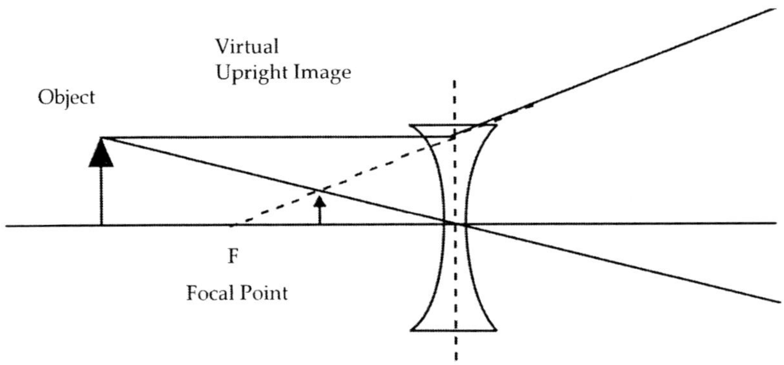

Fig. 9.7 How a concave lens forms a virtual, upright and de-magnified image.

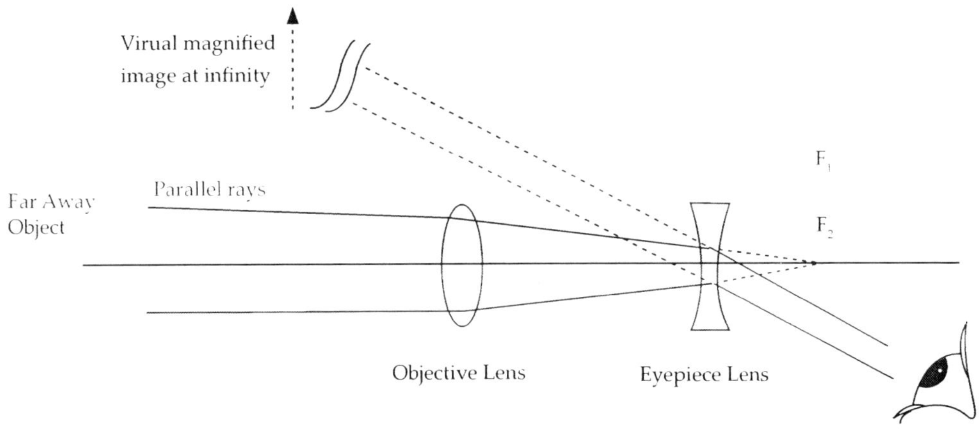

Fig. 9.8: Ray tracing principles of Galileo's telescope.

Under Kepler's treatment, the ancient mysteries of vision, the secrets of the magical optical illusions, the old explanations of reflection and refraction - all dissolved into a precise geometrical and mathematical theory. With keen geometrical analysis, Kepler explained why an object seen in a mirror appears to be placed behind it (Fig. 9.9). The eye that receives the rays from the object after reflection by the mirror has no way of perceiving the actual path followed by the rays. It places the objects as though the rays come straight from it, from behind the mirror. Similarly, he analyzed the displaced appearance of objects by refraction. A stick when thrust into a bowl of water appears to be broken (Fig. 9.10).

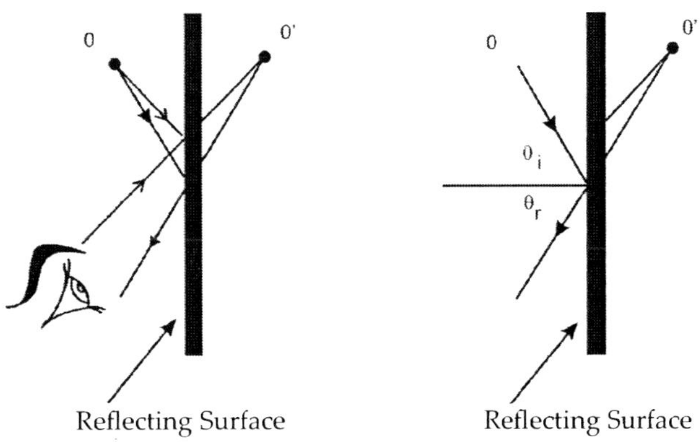

Fig. 9.9: Reflection: the light rays from an object O strike the mirror at an angle of incidence (θi) and reflect off the mirror at angle of reflection (θr) According to the law of reflection, the angle of incidence equals the angle of reflection: θi = θr. The eye-brain thinks that the light is coming from the point O' which is the image of O in the mirror. The image is called virtual because there is no real object at O'.

Kepler suggested an improved telescope over Galileo's version. He proposed a device with two convex lenses, one for the objective and the other for the eyepiece (Fig. 9.11). When an object is located beyond the focal point of the objective, a real image is formed. A second convex lens - the eyepiece - can be used as a magnifying glass to enlarge he real image. Even though the image is inverted, and the telescope length is greater than for Galileo's version, Kepler's telescope is more powerful, and has a larger field of view that remains in sharp focus. Because of these superior features, Galileo' version, which used one concave and one convex lens, evolved only as far as an opera-glass for amusement, while Kepler's version eventually became the astronomer's main research tool.

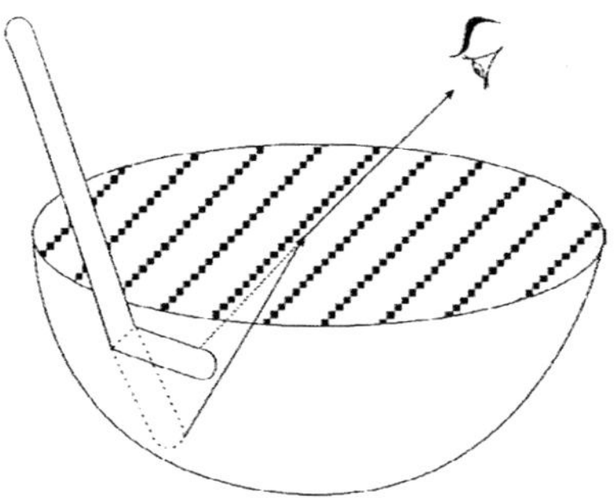

Fig. 9.10: Refraction makes a stick appear bent when immersed under water. A ray of light emerging from the stick under water bends away from the normal at the water-air interface. To the eye it appears as if the light comes from an elevated point - therefore the stick appears bent.

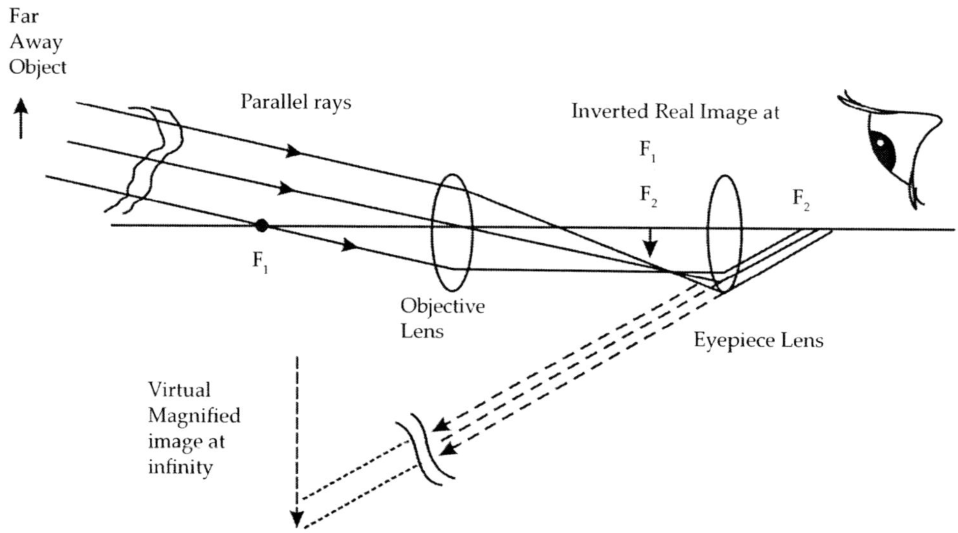

Fig. 9.11 Ray diagrams for Kepler's telescope idea using convex lenses for both the objective and the eyepiece. Parallel rays from a distant object are focused with an objective convex lens to a real image at the focal plane (F1), passing through the focal point of the objective. The eyepiece is also a convex lens placed so that the real image is at the focal plane (F2) of the eyepiece. The real image acts as an object for the eyepiece. The eyepiece forms a magnified virtual image at infinity. The parallel rays are drawn at a different angle for clarity. The magnification turns out to be equal to the ratio of the two focal lengths. The image is *inverted*, but that does not present any problem for astronomical studies.

Chapter Questions

1. Discuss one of the amusing and one of the useful features of lenses that stimulated the evolution of optical science.

2. List two of the scientific and two of the practical advances that emerged from the discovery of the telescope.

3. Describe the difference between the functions and uses of convex and concave lenses?

4. What is the diamond ring eclipse?

5. How did Galileo know that the moons of Jupiter are not stars? Why could they not be some of the many new stars that he could see through his telescopes?

6. Are the moons of Jupiter visible to the naked eye?

7. Do the orbits of the moons of Jupiter obey Kepler's laws? Is there any difference in the expression of the laws?

8. What astronomical observations did Galileo use to support the heliocentric theory?

9. What logical inconsistency did Galileo point out about the Ptolemaic arrangement of the heavenly bodies?

10. Describe in one sentence <u>each</u> how the discovery of <u>each</u> of the following upset Aristotle's picture of the cosmos?

    a. sun spots
    b. nova
    c. moons of Jupiter
    d. the phases of Venus
    e. mountains on the moon.

11. What predictions of Copernicus did Galileo fulfill?

12. Describe the appearance (or phase) of Venus through a Galilean telescope when it first appears as the morning star.

13. What arguments did Galileo give to support the rotation of the earth?

14. Why could the sunspots not be due to orbiting planets, as one of Galileo's contemporaries suggested?

15. How did Galileo know that the sun rotates?

16. What is the effect of the atmosphere in determining the position of the stars?

17. What are some of the differences between the planets and the stars?

18. List three separate effects that dispute the immutability of the heavens.

19. Why does the earth appear to be in a fixed place when viewed from the moon, as opposed to the moon which we see move across the sky every night? Does the earth appear to stand still, or does it rotate?

20. Give three arguments for the non-existence of the celestial sphere of fixed stars.

21. Why was there so much reluctance to accept the new revelations from the telescope?

22. Comment on the following objection to Galileo's discovery of mountains on the moon:

    If the moon is made of the same stuff as the earth, then the moon must have weight. So why does the moon not fall to the earth?

23. What did Galileo discover about Saturn through the telescope?

24. Herschel classified nebulae into several categories. List them.

25. What is the reason that the Milky Way galaxy appears as a belt and not a spherical glow?

26. What is the reason that the planetary movements are confined to constellations in the zodiac belt?

27. Why do we not see stars all way across the Milky Way galaxy?

28. How did Hubble establish that Andromeda is a galaxy outside the Milky Way?

29. How did Shapley determine the true extent of our galaxy even though astronomers could not see stars all the way across the plane of the Milky Way?

30. What is the difference between a open star cluster and a globular cluster?

31. How did Shapley know that the sun is not at the center of the Milky Way?

32. Cite one common law that governs the motion of stars in Milky Way and the planets in the solar system.

33. Discuss some of the new discoveries with the advancing telescopes, and the impact of these discoveries.

Math Based Questions

1. Draw ray diagrams similar to Fig. 9.1 to show how a concave lens will diverge a beam of rays parallel to the axis.
2. Using the table provided, verify Kepler's Laws for the moon's of Jupiter.
3. Using the parallax angle to the nearest star (not the sun) due to the orbital motion of the earth, show why even a telescope with 30 magnification will not be good enough to resolve annual parallax.
4. Draw a diagram of a man standing in front of a mirror that is half his height. Locate the top of the mirror at the same height as the top of the man's head. Draw separate ray diagrams for his head and his feet to show that the man will be able to see his *full* image in the mirror even though the mirror is only half his size.
5. Using the principles illustrated in Fig. 9.6, draw a ray diagram showing how a convex lens can be used as a magnifying glass to aid in reading letters on a page held close to the lens.
6. Draw a ray diagram to show how a concave lens can be used to correct the eye of a near-sighted person.
7. If the sun were to be as far as the nearest star, estimate the angular diameter of the sun as viewed from the earth.
8. Draw a diagram of the earth, moon and the sun to explain how earthshine can illuminate the unlit side of the newly crescent moon. Is it possible to explain earthshine in an earth-centric system?
9. Estimate the relative sizes of the earth and the sun as viewed from the surface of the moon.
10. Suppose we take two identical converging lens and place them next to each other. What happens to the focal length of the combination?
11. What happens to the focal length of a glass lens when it is placed in water? Explain your answer with diagrams.
12. Using the modern value for the speed of light, calculate the time it takes the light from the sun to reach the earth.
13. Using the absolute value of Mars' orbit, and Copernicus' table, determine the diameter of the orbit of Jupiter around the sun.

    Calculate the ratio of the times for thunder and lightning to reach the surface of the earth, if an atmospheric discharge takes place at a height of 1000 meter.
14. Which planet has the shortest period of rotation? How long does it take to complete one rotation?
15. Which planet has the longest period of rotation? How long does it take to complete one rotation?

Exploration Topics

Discuss the many advantages of a reflecting telescope over a refracting telescope. After Newton's invention, why did the reflector not immediately supplant the refractor?

The Best of the Hubble Space Telescope Images

Galileo first turned the telescope to the sky to find new features about the heavenly bodies. The most powerful telescope today is the Hubble Space Telescope (HST). The world wide web site below carries images from the HST and brings them to your computer. In the index you will see three categories (columns) of images: solar system, galactic, extra-galactic. Visit the Hubble site below to learn about some of the more recent discoveries. Treat these as fourth column,

http://hubblesite.org/newscenter/archive/

Spend about one hour looking through images and reading the accompanying text.

Print out one image of your choice from each of the four columns and hand it in.

http://www.seds.org/hst/

1. Write one paragraph of your reflections about HST images.
2. Should the US try to keep Hubble operational or let it fail due to lack of funds? Discuss your answer.
3. How will the new space based telescope under planning be an improvement over the HST?

Sample Quiz

1. Which of the following is true for a convex lens?
   a. It converges rays of parallel light to a real focus.
   b. It can be used as a magnifying glass.
   c. It can be used to focus sunlight to burn a piece of paper.
   d. It can be used to correct the vision of an older person who cannot read very well.
   e. It can be used as an objective lens for a telescope.
2. Which of the following features did Galileo identify about the moon through his telescope and his interpretations?
   a. The moon has mountains.
   b. The boundary line between the dark and the light regions of a crescent moon is perfectly sharp.
   c. The pale glow illuminating the dark portion of the newly crescent moon is caused by earthshine.
   d. The moon rotates.
3. Which of the following features did Galileo identify about the sun through his telescope and his interpretations?
   a. The sun has spots that appear for long periods of time then dissolve.
   b. The sun rotates.
   c. The sun has mountains.
4. Suppose you look at Venus through a Galilean telescope when it first appears as the morning star. Which of the following are true?
   a. It looks round and fully lit.
   b. It appears crescent with the east side lit up.
   c. It appears crescent with the west side lit up.
5. Which of the following did Galileo discover about stars through the telescope?
   a. They appear the same as with the naked eye, except there are many more.
   b. They appear significantly larger.
   c. They look like disks.
   d. He could resolve the annular parallax of the stars due to the motion of the earth.

6. Which of the following did Galileo discover about planets through the telescope?

   a. They appear the same as with the naked eye, except there are many more.

   b. They appear significantly larger.

   c. They look like disks.

   d. Some of the planets show phases like the moon.

7. How did Galileo deduce that the Sun rotates?

   a. Because the earth rotates.

   b. By watching the location of the sun spots.

   c. Because all the heavenly bodies rotate.

   d. By watching the total eclipse of the sun.

8. Among the following choices, Venus is closest to us when it is in

   a. full phase

   b. half phase

   c. quarter phase

   d. crescent phase.

9. In looking at Jupiter and its moon's through the telescope, why do the number of moons that are visible regularly keep changing night after night?

   a. The moons are eclipsed by Jupiter.

   b. Clouds obscure the moons.

   c. The moons are moving too fast.

   d. None of the above.

Answers to Sample Quiz for Chapter 9

1. a to e
2. a & c
3. b
4. b
5. a
6. b to d
7. b
8. d
9. a

Homework Questions for Chapter 9 (with Sample Answers)

1. Using the principles illustrated in Fig. 9.6 (Study Guide), draw a ray diagram showing how a convex lens can be used as a magnifying glass to aid in reading letters on a page held close to the lens.

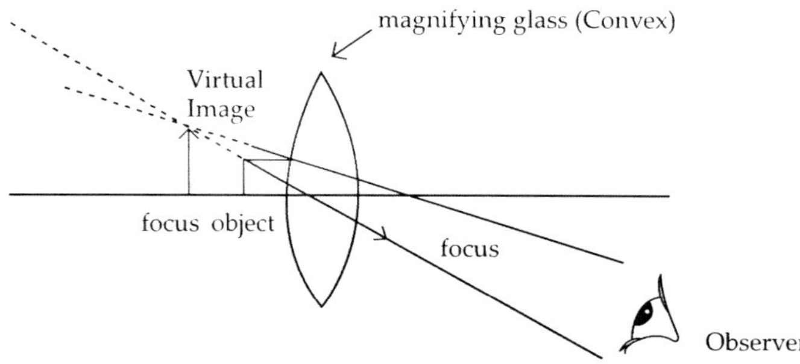

Figure 1: Figure of a magnifying glass. Note that we obtain a virtual, upright and enlarged image on the same side as the object in this case.

2. Draw a diagram of the earth, moon and the sun to explain how earthshine can illuminate the unlit side of the newly crescent moon. Is it possible to explain earthshine in an earth-centric system?

3. Suppose we take two identical converging lenses and place them next to each other. What happens to the focal length of the combination? Draw a ray diagram.

4. Describe the difference between the functions and uses of convex and concave lenses?

5. Describe in one sentence <u>each</u> how the discovery of <u>each</u> of the following Galileo's discoveries upset Aristotle's picture of the cosmos?

    a. sun spots

    b. nova

    c. moons of Jupiter

    d. the phases of Venus

    e. mountains on the moon.

6. What is the reason that the Milky Way galaxy appears as a belt and not a spherical glow?

7. Why do we not see stars all way across the Milky Way galaxy to farthest stars?

    How did Shapley determine the true extent of our galaxy even though astronomers could not see stars all the way across the plane of the Milky Way? How did Shapley know that the sun is not at the center of the Milky Way?

Though it's hard to see far within the plane of the galaxy because of the large density of gas. dust, and stars. Shapely used the many globular clusters of stars that orbit the galaxy in a spherical halo shape, (correctly) assuming that these clusters were distributed uniformly about the center of the galaxy. To determine the distance to these clusters, he identified within them Cepheid variable stars. He used the relationship between the periods of brightness oscillations and the overall brightness of Cepheid variables discovered by Levitt, along with the luminosity distance method, to determine the distances to the clusters. He observed that the great majority of these clusters lie to one side of our field of view, thus showing that we are far from the center of the galaxy.

CHAPTER 10

Reading Questions

P. 522: How did the new paradigm of Descartes' mechanical universe fill the void left over after the demolition of the crystal spherical architecture of the universe?

P. 523: What important astronomical discovery was made by Magellan's crew in the southern hemisphere?

How did this discovery play a role in a later dramatic southern hemisphere find?

How did explorers determine the local time of noon at sea?

How is the longitude difference between two places on earth determined?

Why did it become important to keep track of time as accurately as possible (chronometry)?

P. 524: What were the advantages of mechanical clocks (such as water powered clocks) over sundials?

Why did European clock-makers replace water power to drive clocks with a falling weight?

What is the function of an escapement in a mechanical clock?

P. 526: What was the impact of Galileo's pendulum principle on accurate clock-making?

Describe Descartes picture for the mechanical universe.

P. 527: How was the mechanical paradigm a favorable development for progress in modeling the universe's operations?

P. 528: What were the failures of Descartes' vortex description of the mechanical universe?

P. 530:How did Descartes introduce rational order into the description of space?

P. 531:How did rational order and structure play a role underlying the Baroque period of movement, extravagance and mystic zeal?

Find examples from Baroque art, architecture and music.

Explore the origins of the well-tempered musical scale during this period and its relationships with Pythagorean harmonies.

P. 532:How was the random flow of time controlled, measured and regulated?

P. 533:Examine the role of the beat in regulating music.

What is simple harmonic motion?

What is the connection between circular motion and simple harmonic motion?

P. 535:What is the difference between speed and velocity?

How did Descartes extend Galileo's principle of inertia from speed to velocity?

What was Descartes' key realization about circular motion?

P. 536:Why did scientific activity shift from Italy to France and Holland?

P. 537:What provides the acceleration for the circular motion of a stone whirling on a string on a flat table?

What will eventually happen to the string if the speed of circular motion of a stone on the string keeps increasing?

Why did King Louis XIV provide funding for the progress of science, such as the construction of the Paris Observatory?

How did Huygens advance understanding of circular motion?

P. 538:How did Huygens advance understanding of pendulum motion?

How can you use pendulum swings to measure the value of g?

P. 539:What were the problems of using a pendulum as a clock regulator on ocean voyages?

How did Huygens try to correct some of these problems?

P. 540:Why did the Cayenne expedition astronomers have to change the length of the pendulum to keep their clocks regulated to obtain simultaneous parallax angle measurements for Mars?

P. 541:What provides the centripetal acceleration for the circular motion of objects on a rotating earth?

P. 542:How accurate did Huygens succeed in making his pendulum-regulated clocks?

How did Galileo's pendulum allow Huygens to prove that the earth rotates?

P. 543:Why did it become imperative for maritime powers to have a clock that could keep time to better than 4 seconds per day?

What were the problems with the various astronomical methods proposed to keep accurate track of time and longitude?

P. 544:What were the problems with Galileo's method based on a massive table of eclipse times for Jupiter's moons?

P. 545:What is the advantage of a spring-regulated clock over a pendulum regulated clock?

P. 546:How did Harrison overcome the problems arising from temperature change for the spiral-spring regulated clocks?

What is the conceptual difference between a sidereal and a solar day? What is the actual time difference? Which one is longer?

P. 547:How did Harrison use the time difference to calibrate his new more accurate clocks?

How does the time difference between sidereal and solar days provide proofs for both the rotation and revolution of the earth?

Why did Parliament hesitate for a long time before awarding Harrison the longitude prize?

P. 548:Find out how accurate are the best clocks of today's technology.

P. 549:What is a synchronous satellite?

P. 550:How does Special Relativity play a role in the accuracy of a clock on the GPS satellite relative to an earth-based clock?

P. 551:How will the co-ordinates of a point (p,q,r) change if the origin of the system is moved from (0,0,0) to (a,b.c)?

How will the space-time co-ordinates of an event taking place at (p,q,r,T) change when observed from a vessel moving speed v along the x-axis?

How are space and time linked together for moving observers?

P. 552:Why does a fast moving box not appear the same shape as a stationary box?

At a time when the face of a moving box arrives so that its center is directly in line with a stationary observer, why can the observer see both the face and the side of the moving box? (This would not be possible for a stationary box.)

P. 553:If an alien on planet 70 million light years away from earth were to observe earth at this instant of time, would he/she dinosaurs?

P. 554:Besides seeing the side of the box how would the face of a fast moving box appear different to a stationary observer?

P. 555:Review how chronometry is tied up with circular motion, simple harmonic motion, rotation of earth, and revolution of earth.

Math Topics

Latitude

Example

Suppose, at mid-day, the elevation of the sun on the day of the summer solstice is 83 degrees. What is the latitude of the location?

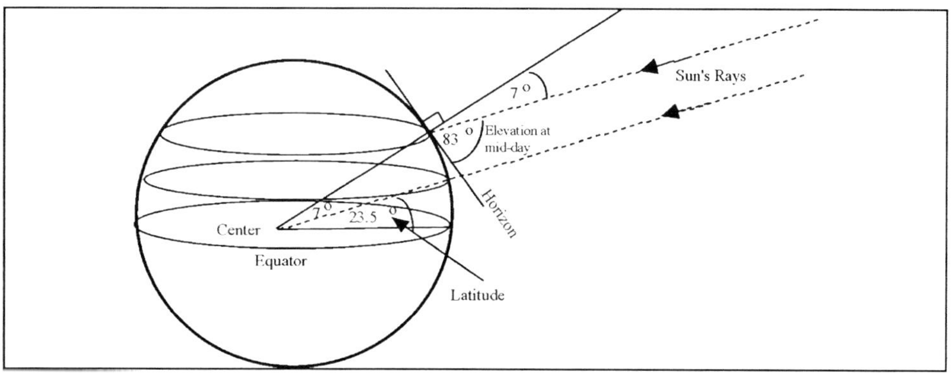

Fig.10.1 The relation between the altitude of the sun and the latitude.

According to the above diagram (Fig. 10.1), at noon on the day of the summer solstice, at a latitude of 23.5 degrees, the sun is at 90 degrees elevation, because of the inclination of the ecliptic with respect to the equator. Recall that at this time the sun casts no shadow at the location of 23.5 degrees (e.g., Syenne). If, at a higher altitude (e.g., at Alexandria), the maximum elevation of the sun on the same day is 83 degrees, then the angle between the radius and the sun's rays is 7 degrees. Therefore the location is 7 degrees further north, i.e., the latitude is 30.5 degrees.

Example

Traveling at a fixed latitude, e.g., along the equator, what is the distance covered if the time difference between two locations is 1 hour.

For every hour time difference, the distance covered will be (1/24) x circumference of the earth, since the ship is traveling along the equator, a great circle around the globe.

$$\frac{1}{24} \times 40{,}000 = 1667 \text{ km}$$

Note that, at a different latitude, one has to use the smaller circumference of the latitude circle instead.

Vectors

Velocity and acceleration are physical quantities with two separate attributes: *magnitude* and *direction*. This is in contrast with *scalar* physical quantities such as time, volume, density and temperature, which do not point in any direction in space, and which can be described by a *single* number, e.g. 4 seconds.

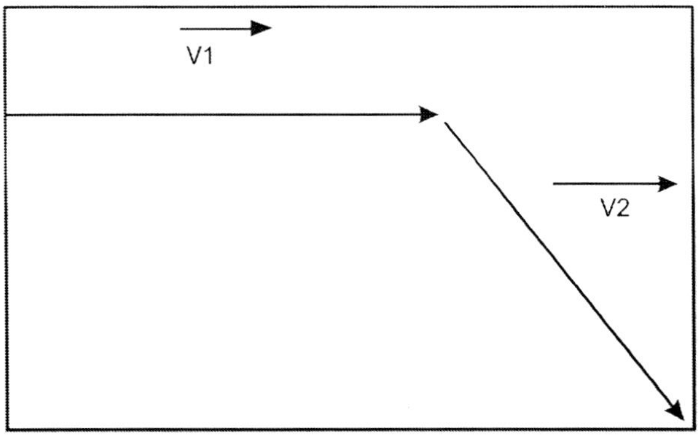

Fig. 10.2 The vector velocity of a body changes abruptly when the direction changes.

In Fig. 10.2, even though the *speed* of the body is the same, i.e., the magnitude of vector $\vec{\mathbf{v}}_1$ equals the magnitude of vector $\vec{\mathbf{v}}_2$.

$$v_1 = v_2$$

but the *velocity* is different. Vector $\mathbf{v}_1$ is not equal to vector $\mathbf{v}_2$. We express vector quantities by putting an arrow on top of the representing symbol, or by making the symbol bold. The magnitude of the vector is written as $V = |\mathbf{V}|$.

Besides velocity, there are many other physical quantities that behave as vectors. Displacement is the most obvious one, appropriate to navigation. If you sail from a point O to the point A in a plane, your displacement vector is (Fig. 10.3)

$$\overrightarrow{OA}$$

If you sail from the point O to the point B, or from the point O to the point C, or from the point O to the point D, your displacement vectors are different in each case. They are:

$$\overline{OB}, \overline{OC} \text{ and } \overline{OD}$$

Each displacement vector has both *magnitude and direction*, but the point of origin does not have to be the same; it can be anywhere in space. The re-positioned displacement vectors shown in figure 10.3 (b) are the same mathematical entities as the ones in (a), even though the origination point of each vector is different.

$$\overline{OA} = \overline{XA} \text{ and } \overline{OB} = \overline{ZB} \text{ etc}$$

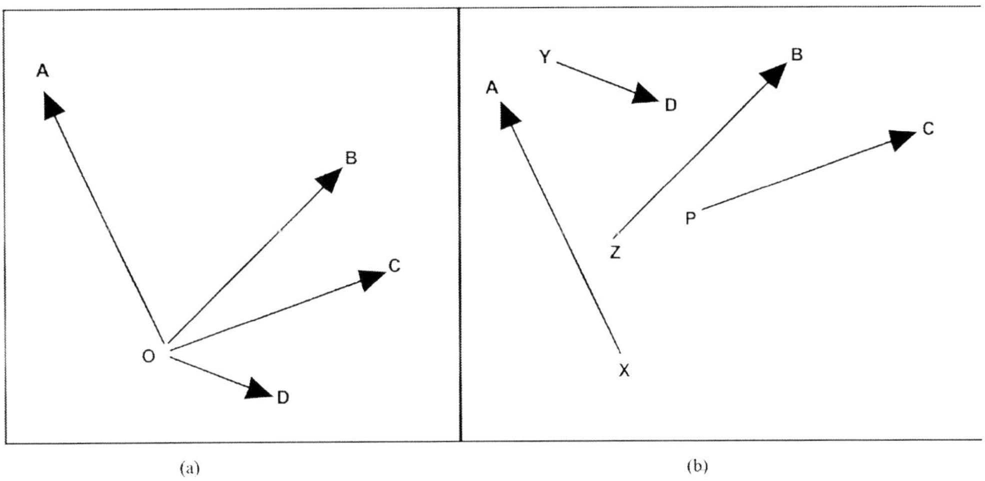

Fig.10.3 Displacement vectors (a) with a *common point* of origin, O. (b) The same vectors, but with different points of origin.

Vector Addition

Vector quantities obey mathematical rules. There is a consistent meaning to vector addition and subtraction, and well defined methods to add vectors, and to subtract vectors. There are also meanings and ways to multiply vectors, but we will not cover those here. From now on, we write *vector* $\overrightarrow{OA}$ as **OA** using the bold and skipping the arrow.

The principles of addition and subtraction of vectors can be illustrated easily using displacement vectors. In Fig. 10.4, you can travel to the final destination C in two ways. In the first route, you can take displacement vector **AB,** followed by displacement vector **BC.** Or, you can directly take displacement vector **AC**. The end result is the same. Therefore in vector addition:

$$\mathbf{AB} + \mathbf{BC} = \mathbf{AC}$$

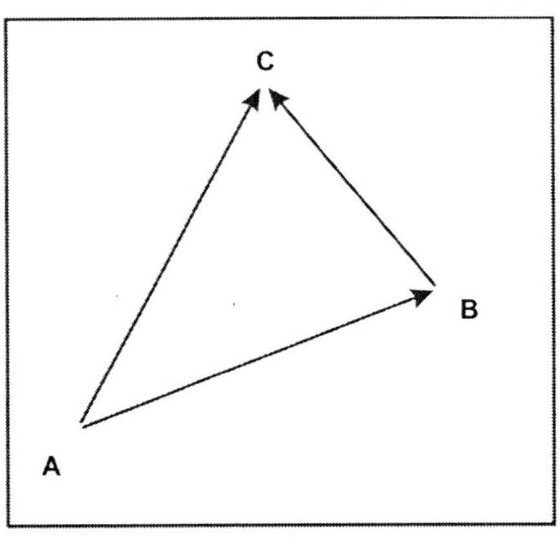

Fig.10.4 Addition of displacement vectors.

AB and **BC** form the two legs of a vector triangle. The sum, also called the resultant, is the third leg of the triangle. This is called the *triangle rule of addition of vectors*.

Like number addition, vector addition obeys the law of commutation.

X + Y= Y + X   number addition

AB + **BC** = **BC** + **AB**   vector addition

Multiplication by a Scalar

Vectors can be multiplied by a scalar (number). For example if **AB** is a vector, then 2 **AB**, 3 **AB** are also vectors, where the direction is the same, but the magnitude is multiplied by 2, 3 ...(Fig. 10.5)

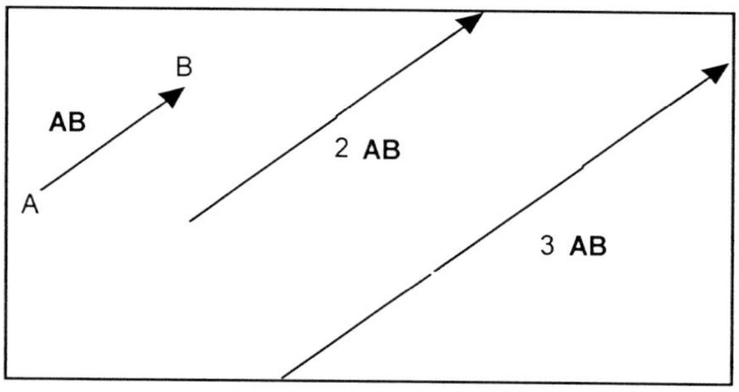

Fig. 10.5 Multiplication of a vector by a scalar.

Vector Subtraction

If you multiply a vector by a *negative number* then the magnitude is the same, but the direction is reversed (Fig. 10.6).

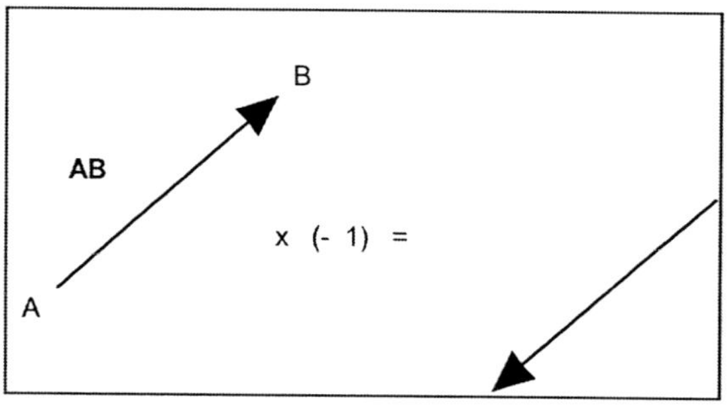

Fig. 10.6 Multiplying a vector by the scalar (-1) reverses its direction.

Now we have a meaning for vector subtraction in terms of vector addition. To subtract a vector, we reverse its direction and add the inverted vector.

AB - **BC** = **AB** + (-**BC**)

Resolving Vectors

Descartes ideas to organize space play an important role in working with vectors. Just as a number can be "broken down" into smaller numbers, which when added together, give the original number,

$$\text{e.g. } 10 = 6 + 4, \text{ or}$$
$$10 = 7 + 3$$

vectors can also be decomposed into component vectors, which when summed together according to the rules of vector addition, yield the original vector. An important way to decompose a vector is along the x- and y- axes of the Cartesian co-ordinate system (Fig. 10.7).

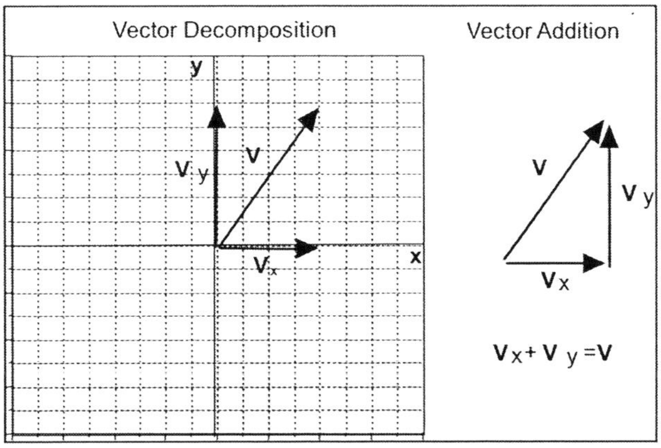

Fig. 10.7 Decomposition of a vector into its rectangular components.

We introduced the definitions of velocity and acceleration in Chapter 2. Now we add directionality. Velocity is also a vector quantity. Since acceleration is defined in terms of change of velocity (as discussed in Chapter 2), *acceleration is also a vector quantity.*

$$Acceleration = \frac{V_f - V_i}{t_f - t_i}$$

Armed with an understanding of velocity as a vector and the new techniques for the mathematical treatment of vectors, we are ready to derive the celebrated expression for acceleration in circular motion, as determined by Huygens. First, we determine the direction of the acceleration vector. In the next step, we see how the familiar geometrical analysis based on similar triangles continues to play an important role in deriving the magnitude of the acceleration. Recall how Thales initiated the powerful geometrical analysis of similar triangles to measure the distance of sailing ships. Later, Aristarchus used similar triangle geometry to estimate the distance from the earth to the moon, and to the sun, as well as to estimate the size of the sun and moon relative to the size of the earth. Now we see how Huygens used similar triangles to derive the magnitude of the acceleration for circular motion.

An object moves with uniform speed in a circle. At time, t= 0, its velocity vector is $\mathbf{V}_i$. At a slightly later time, $\Delta\mathbf{t}$, the velocity changes to $\mathbf{V}_f$. The change in velocity $\Delta V = V_f - V_i$. The acceleration is

$$a = \frac{\Delta \mathrm{V}}{\Delta \mathrm{t}} = \frac{\mathrm{v}_f - \mathrm{v}_i}{\Delta \mathrm{t}}$$

If we subtract the two vectors using the triangle rule that, we see that the direction of the vector $\Delta\mathbf{V}$ is toward the center of the circle (Fig. 10.8). The acceleration of an object in circular motion is therefore *centripetal*, meaning pointing toward the center.

To derive the magnitude of the acceleration, we use two similar triangles in Fig. 10.9, one formed by the procedure of velocity subtraction, $\Delta\mathbf{V} = \mathbf{V}_f - \mathbf{V}_i$, and the other formed by the radius lines and the chord between the two points of the circle, A and B.

Analysis of Circular Motion

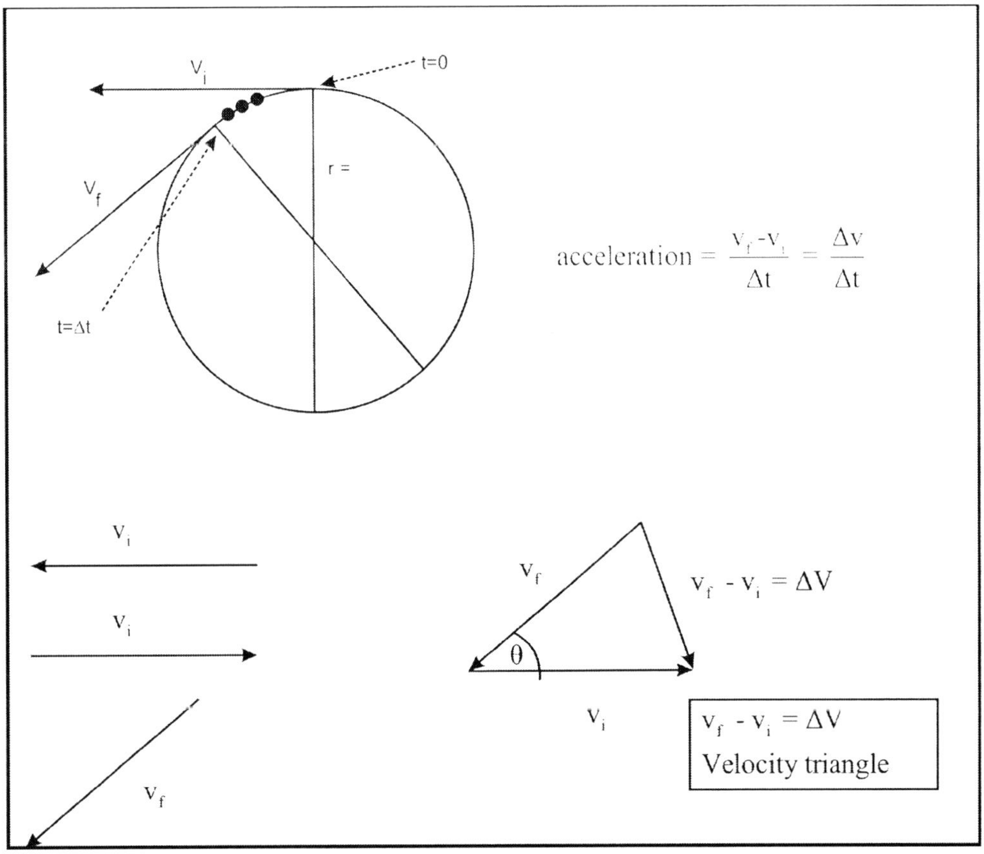

Fig. 10.8 Determining the direction of the acceleration vector. When a body moves in a circle, the magnitude of the velocity vector stays the same, but the direction changes continuously. Consider the velocity vector at two locations. Call these V_f and V_i. The difference is ΔV. We see that ΔV points to the center of the circle. This remains true independent of where we chose to

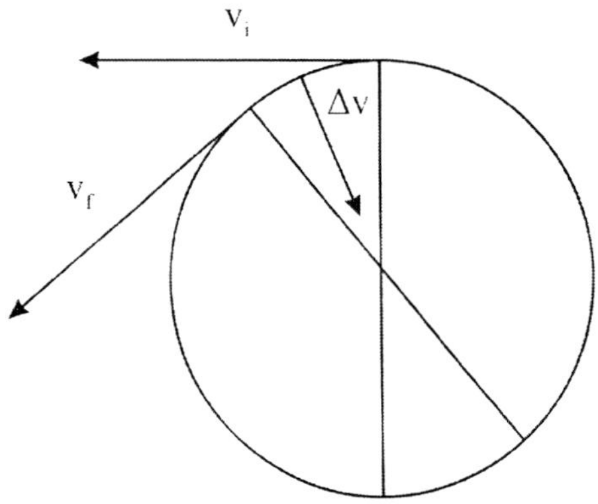

ΔV points towrds center of circle =>
acceleration points towards center

$|\Delta V| = \Delta V$

$$\text{Acceleration} = \frac{\Delta V}{\Delta t}$$

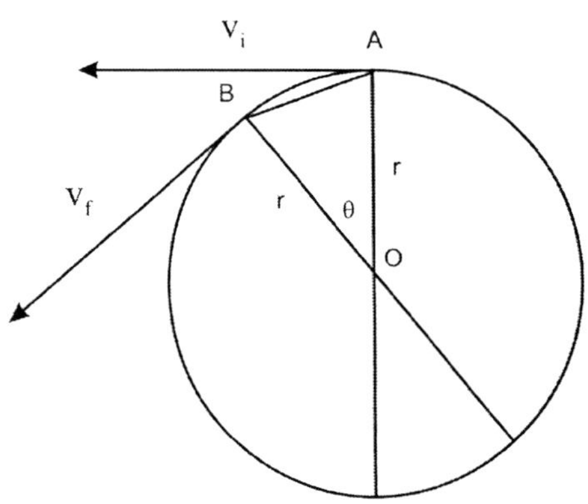

$V_f - V_i = \Delta V$
Velocity triangle

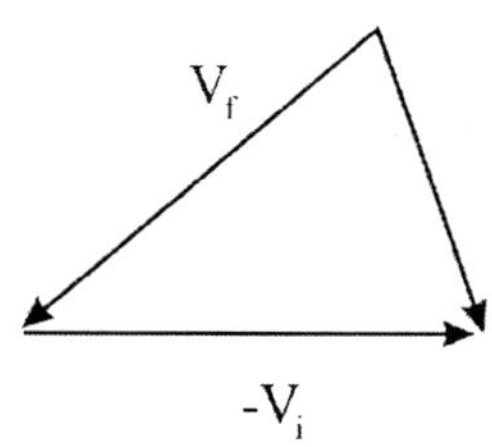

$V_f - V_i = \Delta V$

$|V_i| = |V_f| = V$

Triangle ABO and the velocity triangle are similar

There $\dfrac{\Delta V}{V} = \dfrac{AB}{r}$ $\quad \Delta V = V\dfrac{AB}{r}$

Divide by $\Delta t = \dfrac{\Delta V}{\Delta t} = \dfrac{V}{r}\dfrac{AB}{\Delta t}$

Use $V = \dfrac{AB}{\Delta t}$

$$\text{Acceletation} = \frac{\Delta V}{\Delta t} = \frac{V}{r}V = \frac{V^2}{r}$$

Fig. 10.9 Deriving the magnitude of the acceleration in circular motion.

Simple Harmonic Motion

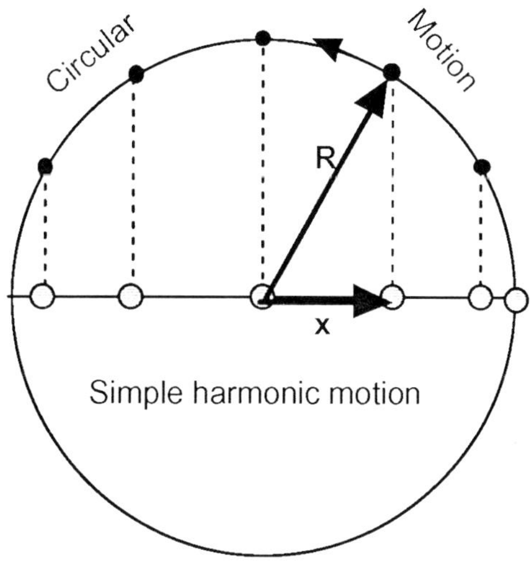

Fig. 10.10 Resolving the position vector, R, of the particle in circular motion gives the displacement vector, x, for the particle in simple harmonic motion.

The particle in circular motion covers equal angles in equal time intervals at a uniform angular speed, for e.g., 10 degrees/sec. The orbital speed of the circular motion particle is

$$V = \frac{\text{distance coverd}}{\text{time taken}} = \frac{\text{cricumferenceofcircle}}{\text{Period}} = \frac{2\pi R}{T}$$

The magnitude of the acceleration of the circular motion particle is

$$\frac{V^2}{R} = \frac{1}{R} = \frac{(2\pi R)^2}{T} = \frac{1}{R}\frac{4\pi R^2}{T^2} = \frac{4\pi^2 R}{T^2}$$

As we already discussed, the direction of the acceleration vector is toward the center of the circle. We now resolve the acceleration vector into its Cartesian components to find two important features about the acceleration of the particle executing SHM (Fig.10.10)

$$a = -\frac{4\pi^2 R}{T^2}$$

$$a_x = -\frac{4\pi^2 X}{T^2} \text{ (x component of acceleration)}$$

1. The magnitude of the acceleration is proportional to the position of the particle. The further out the particle is from the origin, the larger the acceleration that tends to restore it to its equilibrium position, which is the center, or the origin.

2. The acceleration vector is directed towards the origin.

These are the fundamental properties that distinguish *simple harmonic motion*. If there is any form of cyclic motion that follows these basic relationships between the acceleration and the position, the proportionality constant of this expression -($\frac{4\pi^2}{T^2}$) - can be used to determine the period of the oscillation, T.

In particular, we are interested in the relationship between the acceleration and the position of a pendulum. The downward, free fall acceleration vector **g** can be resolved into two components, one along the string, and the other perpendicular to the string. Downward motion along the string is prevented by presence of the string, i.e., by the tension in the string. But acceleration perpendicular to the string causes SHM oscillations of the pendulum bob. The oscillations of the pendulum bob are caused by a component of the free fall acceleration, as Galileo intuitively recognized when he drew important lessons about free fall motion from the swinging pendulum.

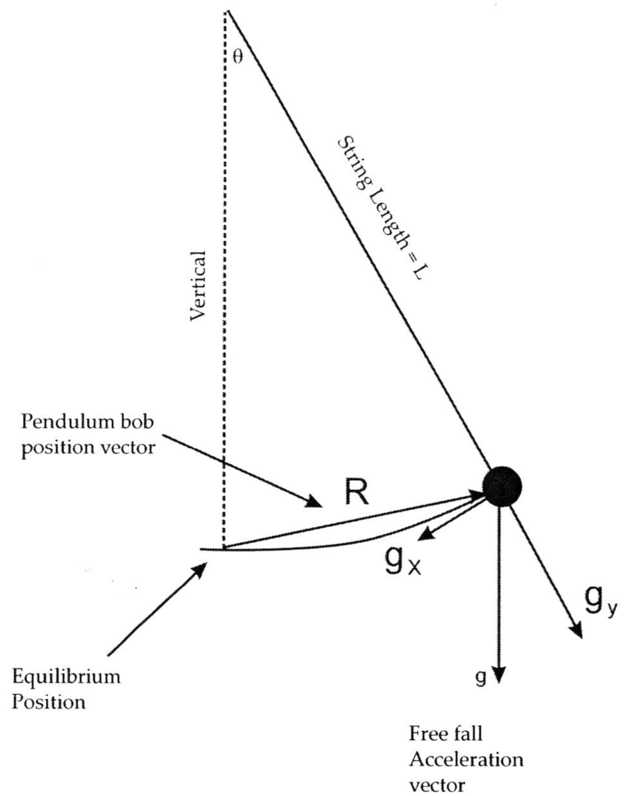

Fig. 10.11 Resolving the acceleration vector, g, of the pendulum bob. g_x is responsible for accelerated motion. The tension in the string prevents any motion in the direction of the string.

Similar Triangles for Analysis of the Pendulum

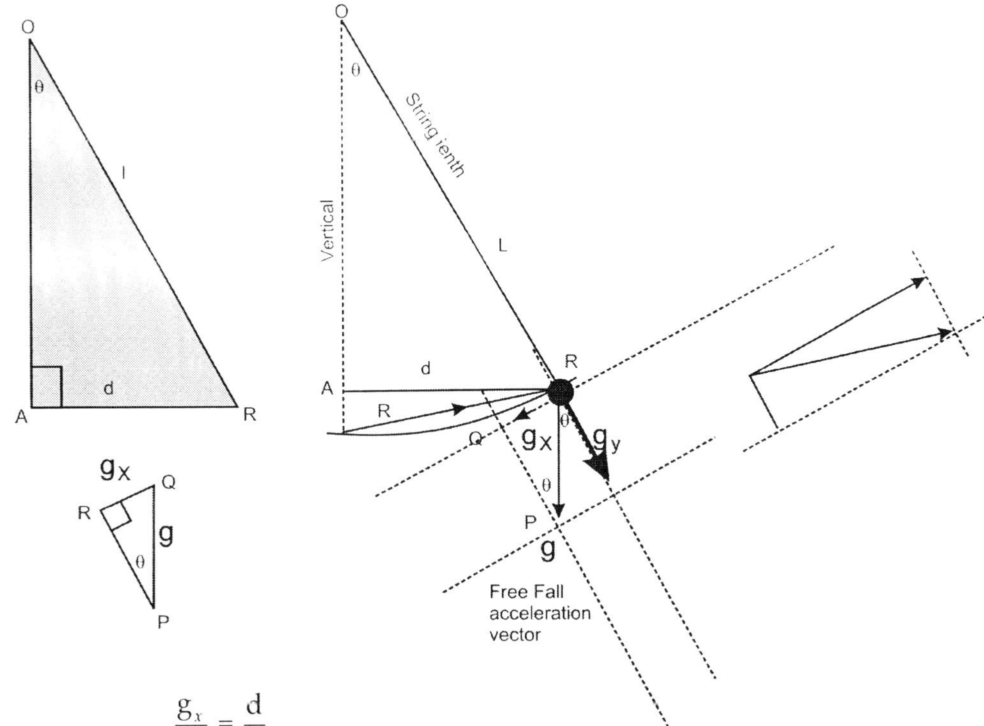

$$\frac{g_x}{g} = \frac{d}{l}$$

$$g_x = g\frac{d}{l}$$

Fig. 10.12: Similar triangles, OAR and PQR are used to calculate the acceleration of the pendulum bob.

Once again, using similar triangles, as in Fig. 10.12, we can relate the acceleration magnitudes (g_x and g) to the important distances, i.e., to the excursion (d) of the pendulum bob, and the length (L) of the string.

$$\frac{g_x}{g} = \frac{d}{l} \quad g_x = g\frac{d}{L}$$

To progress, we need to make an *approximation*. The fact that we are making an approximation should not be interpreted as though we are about to derive an inaccurate relationship. Following Galileo's example, we aim to gain one step of understanding at a time. For small excursions of the pendulum swing, the value of the angle θ is small, and d << L. Under this condition, the distance traveled by bob along the x-direction is also equal to d,

$$d \approx |x|$$

where **x** is the x-component of the vector R, the position vector of the pendulum bob. Our approximation is valid over a rather large angle. Further analysis shows that if the swing angle is less than 23 degrees, the inaccuracy is less than 1%. The key approximation then allows us to link the acceleration with the displacement.

$$gx = g\frac{d}{L} \approx -\frac{g}{L}x$$

Therefore, the acceleration of the pendulum in the x-direction is related to the x-position of the pendulum in the special way that is characteristic of SHM, according to the conditions on page 165. Now that we are confident that the motion of the pendulum is simple harmonic, we can write down the expression for the period of oscillation (T) in terms of the proportionality constant (g/L) and $\frac{4\pi^2}{T^2}$. Therefore,

$$\frac{4\pi^2}{T^2} = \frac{g}{L} = \text{and rearranging, } T = 2\pi\sqrt{\frac{L}{g}}$$

We finally arrive at the famous formula that allows us to calculate the period of a pendulum in terms of its length. Once again we have a specific relationship between space and time. A time period of one second can be cast in terms of the length of a pendulum.

Example

Compute the magnitude of the centripetal acceleration of an object located on the earth's equator.

Radius of the earth at equator = 6378 km

The period of rotation of earth = 24 hour = 86, 400 sec

Compute the tangential velocity, V

$$v = \frac{2\pi R}{T} = \frac{2\pi \times 6378}{86400} = 0.464\frac{km}{sec} = 464\frac{m}{sec} = 1670 = \frac{km}{hr}$$

Note that when converted to mph, the rotation speed of the earth is a surprising 928 mph, more than twice as fast as an airplane! Now we calculate the acceleration.

$$a = \frac{V^2}{R} = \frac{464^2}{6378 \times 1000} = 0.0338\frac{m}{sec^2}$$

Note how small the result is compared to $g = 9.8\frac{m}{sec^2}$

Example

How long does a pendulum have to be to have a complete period of oscillation of one second?

The period of a pendulum is given by

$$T = 2\pi\sqrt{\frac{L}{g}}$$

Therefore

$$L = g\frac{T^2}{4\pi^2}$$

$$L = 9.8\frac{1^2}{4\pi^2} = 0.25 \text{ meter}$$

Example

What is the effect on the period of a pendulum due to the difference in g between the equator and the pole of the Earth?

As we showed in an earlier example, any object at rest on the equator of the spinning earth is already accelerating toward the center of the earth at

$$0.0338\frac{\text{m}}{\text{sec}^2}$$

An object at rest on the pole of the earth would have zero circular motion about the center of the earth, and its centripetal acceleration is therefore zero. This means that the net acceleration of an object on the pole is different from that of an object on the equator by $0.0338\ \frac{\text{m}}{\text{sec}^2}$. If the acceleration due to earth's attraction is 9.8 m/sec^2 at the pole, at the equator the acceleration is 9.8 - 0.0338 m/sec^2 = 9.7662 m/sec^2.

The period of a pendulum is given by

$$T = 2\pi\sqrt{\frac{L}{g}}$$

where L is the length and g the acceleration. If g = 9.8 m/sec^2 at the pole, and L = 1 meter, then from the pendulum formula:

T = 2.0071 sec for one complete oscillation.

But at the equator g= 9.7662 m/sec^2. Therefore, re-applying the pendulum formula,

T = 2.01056 sec at the equator

The percentage error in time measurement is only 0.17%. That seems like a small difference, but over a day (86,400 sec) a pendulum will lose 147 sec = 2.44 minutes! Remember that Huygens had improved the precision of the pendulum to tell time to better than one minute per day. Now his pendulum was sufficiently accurate to detect the earth's rotation.

A pendulum with an unknown 2.5 minute *daily* error would be totally unsatisfactory for keeping track of longitude on a long ocean voyage. At the end of a month-long journey across the Atlantic ocean, the corresponding time reading would be off by 75 minutes and the longitude assessment by 19 degrees! From the radius of the earth, we can calculate that each degree of longitude around the equator is about 112 km, so that the distance error would be 2128 km. At sea, such a distance error would amount to a grievous miscalculation. A ship could arrive at an unfriendly port, or run aground on unexpected shallow rock.

Example

There is a famous old movie called "Around the World in 80 Days". The trip is comparable in time to that of Magellan's circumnavigation of the globe (Chapter 8). Calculate the time it would take to reach Saturn on an imaginary journey in Columbus' ship.

From the circumference of the earth we find that the speed of the vessel is 4×10^4 km/ 80 days = 500 km/day.

The distance from Saturn to the sun is 1.4×10^9 km and the distance from the earth to the sun is 1.5×10^8 km.

Therefore the distance from earth to Saturn (when it is closest to us) is 1.25×10^9 km.

It will take 2.5 million years to reach Saturn - more than the time that humans have inhabited the earth!

Example

Our air travel technology has now advanced to the point that we can circle the globe in one day. At present speeds, calculate the time it would take to reach Saturn.

Airplane solution

86 years - the entire human life span!

Our space technology has advanced to the point that we can circle the globe in one hour. Recalculate the time it would take to reach Saturn.

In the space vehicle we can reach Saturn in 3.6 years.

Is it possible that our capabilities will increase further?

Chapter Questions

1. Discuss two benefits to science from advances in accurate measurement of time intervals. Discuss two immediate practical benefits to society.
2. Why was the sun-dial imported from Greece to Rome unusable?
3. Why is a sun-dial clock useless for navigation?
4. Give two difficulties of the sun-dial that a water-clock resolved.
5. What are the disadvantages of a water-clock?
6. Give three separate reasons why the pendulum could not keep the accurate time needed for longitude measurements?
7. Discuss some of initial difficulties associated with spring-driven clocks as a time keeping mechanism. How did the difficulties with the spring-driven clocks get solved?
8. Why does the value of g change from location to location on the earth's surface?
9. List 10 examples of regular cyclic motion.
10. Which is the better method to determine latitude: from the North Star or from the altitude of the sun?
11. Discuss some of the incentives for exploration that led to sea routes to the East and to the discoveries of new (for the Europeans) continents.
12. Discuss (a) a practical benefit and (b) a scientific advance that resulted from a more precise determination of the value of g.
13. Give two separate methods each for determining (a) latitude and (b) longitude.
14. Discuss how technological advances in time keeping after Galileo influenced specific advances in science.
15. Discuss how the scientific advance of circular motion and simple harmonic motion influenced our ability to keep track of time.
16. Discuss the evidence for the rotation of the earth.
17. Explain with diagrams why altitude of the pole star does not depend on (a) time nor (b) the calendar. It depends only on the latitude.
18. Discuss how time and space are linked in
    a. determining latitude
    b. determining longitude
    c. accurate timekeeping.

19. Apart from the problem of a stable telescope at sea, can you think of another technical difficulty with Galileo's idea to use the eclipses of Jupiter's moon as a clock every night?

20. Give three separate reasons why a pendulum could not be used as an accurate time keeper on ocean voyages for telling longitude.

21. Why is a spiral spring- flywheel based clock better for accurate time keeping than a pendulum clock?

22. Cite three separate contributions of Descartes to the advancement of science.

23. What is the relationship between circular motion and pendulum motion?

24. A car makes a sharp right turn. What kind of motion do the passengers in the back seat experience?

25. What is the direction of acceleration of the cart when it reaches the top of the loop-the-loop portion?

26. How does a three dimensional object look different to two observers from different locations?

27. How does a three dimensional object look different to two observers who arrive at the same location but are traveling at different speeds?

28. How does special relativity play a role in the accuracy of GPS systems used for navigation?

Math Based Questions

1. How many degrees does the moon move daily?

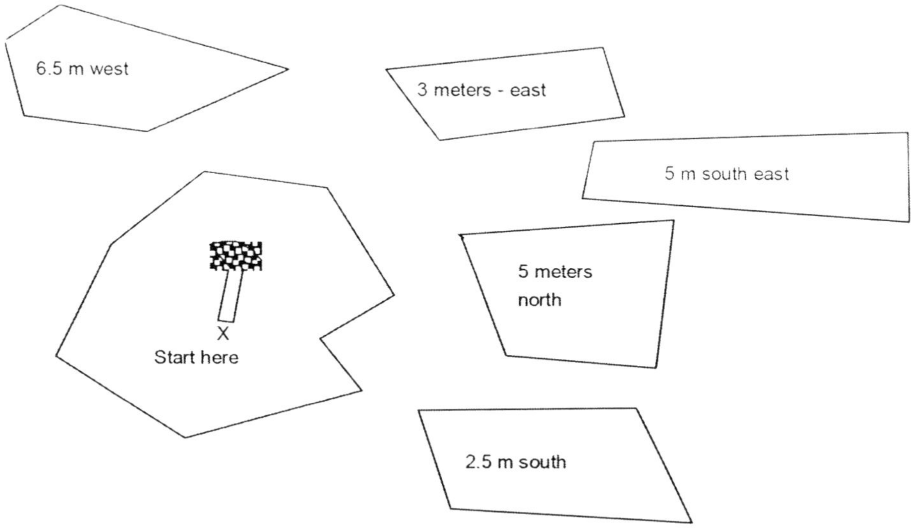

2. A treasure map (above) was found torn into 6 pieces as shown above. One piece showed a tree that was recognizable in the surrounding landscape where the torn map was found. The problem was that there were no clues as to the order in which to follow the directions in the different pieces. Find the location of the treasure relative to the spot X below the tree. Show, by using vectors, that the order in which the steps are taken does not matter. We always get to the same final resting place of the treasure.

3. What is the magnitude of a displacement whose components are along the x-axis 4 m and along the y-axis 2.5 m?

4. An object moving in a circular path with a constant speed of 2m/sec changes direction by 30 degrees in 3 seconds.

   a. What is the change in velocity?

   b. What is the average acceleration during the 3 seconds?

5. A watch has a second hand that is 2 cm long.

   a. Compute speed of the tip of the second hand.

   b. What is the velocity of the tip?

   c. What is the acceleration of the tip?

6. What is the period of a phonograph record that turns at 33 & 1/3 rotations per minute?

7. In the heliocentric system, draw a diagram to show the direction of the acceleration of the planet Venus, the moon of the earth, and a moon of Jupiter.

8. Compute the centripetal acceleration of the earth toward the sun in m/sec^2 (assume a circular orbit).

9. Compute the centripetal acceleration of the earth toward the sun in terms of g (assume a circular orbit).

10. Check the formula for period of the pendulum for units, i.e., carry out a dimensional analysis.

11. At the equator, what is the length of a pendulum with the total to-and-fro period of exactly one second? At the North pole, what is the period of the same length pendulum?

12. How long is a pendulum with a period of 10 sec?

13. Give the formula for the *frequency* of oscillation of a pendulum in terms of g and L.

14. If the length of a pendulum is increased by a factor of 2 what will happen to its period of oscillation?

15. How fast would the earth have to spin so that the net acceleration to the center of the earth is zero? Give your answer in rotations per day.

16. Does the altitude of the North star depend on the seasons? Explain your answer with a diagram.

17. Using Fig. 10.1 as a guide, draw a diagram to show that because of the seasons, a simple sun-dial which is calibrated at one latitude will be totally inaccurate at another.

    The following three questions are based on the Newton's proposed experiment to prove the rotation of the earth, an experiment which Hooke carried out successfully. Drop a stone from a tower 100 meters high. On first consideration, because the tower rotates with the earth, the stone should hit the ground at the base of the tower. This was pointed out by Galileo. But Newton noted that the top of the tower rotates with higher velocity than the base of the tower because the top is at a higher radius. Therefore the stone will fall slightly to the *east* of the base.

18. Calculate the *excess* (above the base) horizontal velocity of the stone at the top of the tower

19. Calculate the time it takes for the stone to hit the ground

20. Calculate how much to the east of the base the stone will land on the ground.

21. A year is a division of time defined by the period of revolution of the earth about the sun. Various kinds of astronomical years and calendar years have been defined. The astronomical year of chief importance is the tropical year, which is the time interval between two successive occurrences of the spring equinox. Its length is now 365.2422 mean solar days, but it decreases very slowly as a result of small, progressive changes in the earth's rotational speed as well as its orbit about the sun. The sidereal year, determined by the earth's position with respect to the stars, is 365.25636 mean solar days. It is longer than the tropical year because it is not subject to the shortening effects of precession. Calculate the difference between the sidereal year and the solar year in minutes.

22. Repeat the graphical vector analysis of Fig. 10.9 for a different position of the orbiting point to show that the change in the velocity vector will again point to the center of the circle.

23. With diagrams based on the heliocentric system, show why the altitude of the zenith sun depends on the month, but the altitude of the pole star does not.

24. Traveling at a fixed latitude of 30 degrees N, what is the distance covered if the time difference is one hour.

25. Draw diagrams to show why the lengths of shadows change with latitude at three different latitudes.

26. What is the ratio of the speed of light to the speed of earth's orbit around the sun?

27. How long does it take light to travel from the sun to the earth?

    from Saturn to the earth?

28. Calculate the period of oscillation of a one meter long pendulum on the poles where $g = 9.8\ m/sec^2$.

29. Calculate the period of oscillation of a one meter long pendulum on at the equator.

    How can the period of a pendulum be used to prove the rotation of the earth?

30. A loop in a roller coaster ride has a radius of 10 meters. Calculate the minimum speed a cart must have so that the people in the cart do not fall off the when they reach the top of the loop and are upside down.

31. How many rotations per day would the earth have to make in order that the centripetal acceleration of an object on the equator would become equal to g?

32. Suppose the earth were in fact rotating at this high speed. A person on the equator releases a ball from his hand. What would the effect of a ball let go from a one meter height?

Exploration Topics

1. Discuss the following statement:

   "The rewards of exploration are sometimes a confirmation of the motivating goal, but more often they open new, unexpected and more exciting vistas."

2. Draw parallels between sea voyages and modern space explorers to support or contest the claim:

   "There seemed to be no limit to the possibility for fresh discoveries and new capabilities."

3. Discuss the statement below in terms of the events of this chapter. Relate it to some of the more recent developments in our culture:

   "Every advance in science and technology is a double edged sword."

4. Write a dialogue between Columbus and Huygens on the discovery of new worlds.

5. Discuss some of the advantages and some of the disadvantages of governmental support for science in terms of the experiences of the French Academy. Do you see any parallels today?

6. Discuss Aristotle's statement: "Not only do we measure time by movement but also movement by time, since they define each other."

7. How far have spacecraft from Earth traveled into space? Give your answer in light years.

Sample Quiz Questions

1. Two golf balls are dropped from the top of a track in the shape of a cycloid as in panel A of the figure below. Which panel represents the result at the time when the first ball reaches the bottom?

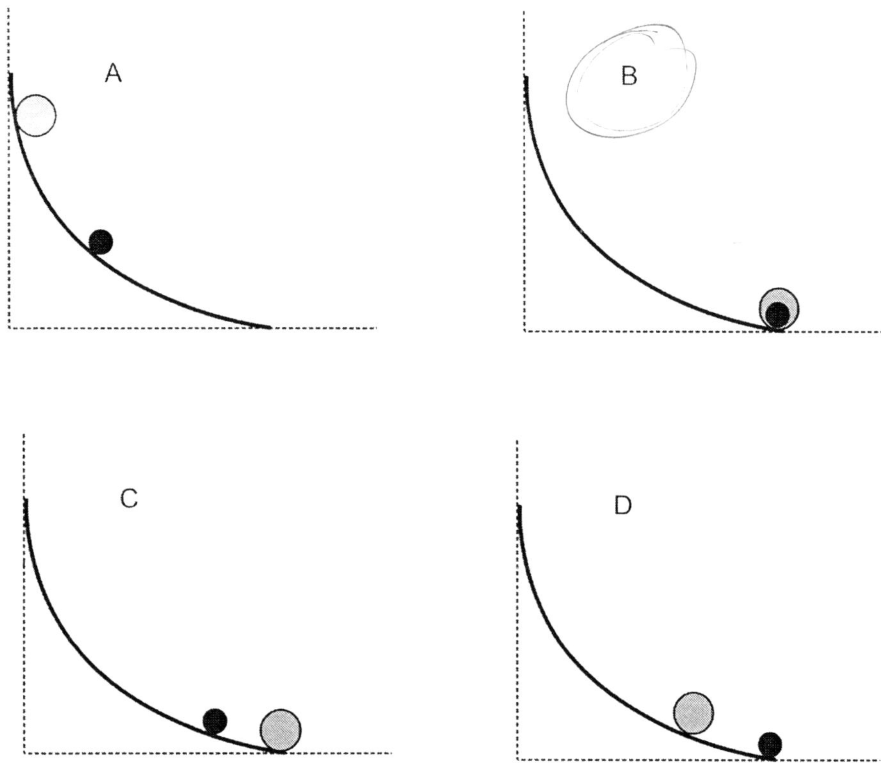

2. Which of the following is appropriate to simple harmonic motion?

   a. It describes the motion of a pendulum.

   b. It describes the motion of a point on a vibrating string that emits a musical tone.

   c. The acceleration is proportional to the displacement and in a direction opposite to the displacement.

   d. All of above.

   e. None of the above.

3. On a loop-the-loop roller coaster ride, the cart and riders at the top of the loop do not fall to the ground in spite of the earth's gravity. This is because the centripetal acceleration becomes equal to the free fall acceleration. If the radius of the circular part of the loop is 10 meters, determine the (approximate) minimum speed of the cart at the top?

   a. 1 m/sec b. 10 m/sec c. 100 m/sec d. 1000 m/sec

4. The period of a pendulum is independent of the amplitude of swing provided

   a. The length of the pendulum is one meter.

   b. The excursion of the swing is small.

   c. The path of the arc follows a cycloid.

   d. The pendulum is not at the north pole.

   e. b. and c.

5. Suppose the length of a pendulum is cut to a quarter of its original length. The period of the pendulum will be

   a. longer by a factor of 2

   b. shorter by a factor of 2

   c. longer by a factor of 4

   d. shorter by a factor of 4.

Which of the following was the most significant problem with the timekeeping capabilities of

6. sun-dial

7. water-clock

8. pendulum regulated clock

   a. The value of g changes with latitude.

   b. The length of the hour markings at one latitude are not valid at another latitude.

   c. The impulse delivered to the clock mechanism would vary with time.

   d. It was difficult to run in the cold winter months.

Identify the best match for each of the following cases of motion

9. A cannon ball falling from the Tower of Pisa.

10. A one kg mass suspended from a spiral spring.

11. Motion of a planet around the sun.

12. A fly sitting on a phonograph record.

   a. simple harmonic motion

   b. constant acceleration

   c. circular motion

   d. parabolic trajectory

   e. elliptical trajectory

13. Why does the value of g change from location to location on the earth's surface?

    a. Because the earth is round.

    b. Because the earth revolves around the sun.

    c. Because the earth's rotation axis is tilted.

    d. Because the earth spins.

    e. None of the above.

14. A watch has a second hand that is 2 cm long. Compute speed of the tip of the second hand.

    a. 0.21 cm/sec b. 2 cm/sec

    c. 20 cm/sec d. 0.021 cm/sec

15. Compute the centripetal acceleration of the earth toward the sun in m/sec^2 (assume a circular orbit). Earth-sun distance = 1.5×10^{11} m

    a. 3×10^4 m/sec^2 b. 21.4 m/sec^2

    c. 6×10^{-3} m/sec^2 d. 3.43 m/sec^2

16. A fielder throws a baseball to the pitcher at a 45 degree angle to the horizontal and at a speed of 20 meters per second. What is the horizontal component of the speed?

    a. 20 m/s b. 10 m/s

    c. 5 m/s d. 14 m/s

17. A swimmer jumps into a river flowing east at 20 mph. If his swimming velocity in still water is 20 mph towards north, what is his resultant speed in the stream?

    a. 28.3 mph b. 20 mph

    c. 14.2 mph d. 40 mph

Answers to Sample Quiz for Chapter 10

1. b
2. d
3. b
4. e
5. b
6. b
7. d
8. a
9. b
10. a
11. e
12. c
13. d
14. a
15. c
16. d
17. a

Homework Questions for Chapter 10 (with sample answers)

1. Give three separate reasons why a pendulum could not be used as an accurate time keeper on ocean voyages for telling longitude.

   Sample Answer

   Sailors would have difficulty using a pendulum clock as a time keeper at sea for several reasons:

   a. The rocking of the ship would interfere with the periodic motion.

   b. Changing temperature and humidity would change the length of the pendulum affecting its period.

   c. The value of g changes with latitude due to the rotation of the Earth. This also affects the pendulum's period.

2. How many rotations per day would the earth have to make in order that the centripetal acceleration of an object on the equator would become equal to g?

3. Suppose the earth were in fact rotating at this high speed. A person on the equator releases a ball from his hand. What would the effect of a ball let go from a one meter height?

4. Calculate the period of oscillation of a one meter long pendulum on the poles where $g = 9.8$ m/sec^2. Calculate the period for a pendulum on the equator.

5. A loop the loop in a roller coaster ride has a radius of 10 meters. Calculate the minimum speed a cart must have so that the people in the cart do not fall off the when they reach the top of the loop and are upside down.

6. What is the direction of acceleration of the cart when it reaches the top of the loop-the-loop portion?

7. Compute the centripetal acceleration of the earth toward the sun in m/sec^2 (assume a circular orbit).

   Compute the centripetal acceleration of the earth toward the sun in terms of g (assume a circular orbit).

   Again, we use the centripetal acceration formula:

$$a = \frac{v^2}{r} = 4\pi^2 \frac{r}{T^2} = 4(3.14)^2 \frac{1.5 \times 10^{11} m}{(3.15 \times 10^7 s)^2} = 5.95 \times 10^{-3} \frac{m}{s^2}$$

   where we have used the expression for the speed of the Earth, $v = 2\pi r/T$, $r = 1.5 \times 10^{11} m$ is the Earth-Sun distance, and $T = 1yr = 3.15 \times 10^7 s$ is the period of the Earth's orbit.

In order to convert to units of g, we multiply by the appropriate conversion factor:

$$a = 5.95 \times 10^{-3} \frac{m}{s^2} \frac{1g}{9.8m/s^2} = 6.08 \times 10^{-4}.$$

CHAPTER 11

Reading Questions

P. 556: How did Newton bring the works of Kepler and Galileo together?

P. 557: What were the parallel and contrasting factors underlying the turmoil that surrounded Newton's growth and Kepler's?

P. 558: Give some examples of Newton's mechanical prowess during his youth.

P. 560: How is Newton's first law of motion different from Galileo's principle of inertia?

How did Newton relate his definition of force to the nature of motion?

How does Newton's force law define the meaning of mass?

P. 561: How does Galileo's law of inertia become a special case of Newton's force law?

How did Newton use the force law to calculate motion?

P. 562: How did Newton use the second law to determine unknown forces acting in motion?

P. 563: What is the source of the force which accelerates the falling apple towards the earth?

What is the magnitude of that force in units of newtons?

How does the earth manage to keep up the same acceleration at the same value g for an apple as for an apple with twice the mass?

What did Newton conclude about the nature of the gravity force from the above behavior: acceleration due to earth's gravity is independent of mass?

If earth's force of gravity extends all the way up to the moon, why does the moon not fall down to earth?

P. 564: How did Newton determine the centripetal acceleration of the moon's circular motion?

Why is the centripetal acceleration of the moon much smaller than g?

P. 565: How did Newton determine that the force of gravity decreases with the square of the distance to the object it acts on?

P. 566: What did Newton conclude about the gravity force that the sun exerts on the earth (and on other planets) from his analysis of the planets (approximate) circular motion and the application of Kepler's 3rd Law?

P. 567: How did Newton realize that the mechanical bond between the sun and the planets was the same gravity force between the earth and the moon, between the earth and the apple?

P. 569: What role did Halley play in persuading Newton to publish his discoveries?

What was the source of the long dispute between Newton and Hooke?

P. 570: What is the essential property of the gravity force behind Kepler's 2nd (area) Law?

P. 572: How does Newton's 3rd Law of motion bring symmetry to the force of gravity between the earth and the moon? Between the sun and the planets?

P. 573: If the apple attracts the earth with a force of the same magnitude as the earth attracts the apple, what happens to the earth's acceleration toward the apple?

How did Newton make the gravity force law symmetric with respect to the masses of the interacting bodies?

P. 574: Why did Newton generalize the law of gravity to any two bodies in the universe?

How did Newton bring symmetry, unity and elegance to his law of gravity?

What new elements beyond Galileo did Newton introduce into his concept for gravity?

P. 575: If the sun attracts the earth with the force of gravity, why does the earth not fall directly into the sun?

How did Newton account for the strange property that the mass in the gravity force law is the same as the mass in the second law of motion?

How did Newton account for the strange property that the force needed to lift an object (against earth's gravity force) increases with the mass that resists the change in its state of motion (inertia)? This question is the same as the one above.

P. 576:How did Newton estimate the mass of the earth?

How did Newton estimate the density of the earth?

P. 577:How did Cavendish confirm Newton's guess for the mass of the earth?

P. 578:What is the relation between Kepler's 3rd Law and the mass of the sun?

How did Newton determine the mass of Jupiter?

P. 579:What would be the duration of a day if the earth were to rotate fast enough that the apple from a tree does not fall to the ground but sits at the same height?

P. 582:Under what conditions would the path of a satellite become elliptical? Hyperbolic?

What prevents the stars from falling into the center of the Milky Way?

P. 584:What calculations did Newton carry out about the 1682 comet and what were the consequences of those calculations?

What observations did Halley make about the 1682 comet, and what were his predictions based on those observations and Newton's work?

P. 585:How did Halley and Newton's work on the 1682 comet change the long-held view about comets?

What was the impact of the return of Halley's comet?

P. 587:What led many to guess that the moon must be responsible for tides ?

Why do high tides at a particular location repeat every 12.8 hours?

Why does the sun play a lesser role in the formation of tides than the moon even though the gravity force of the sun on the earth is much larger?

P. 590:What is the phase of the moon when the tides are strongest (spring tides)?

How often do spring tides occur?

What explanation did Newton offer for the precession of the equinox?

P. 592:What is the connection between precession and the earth's shape?

P. 593:What is the effect of the earth's shape on the period of a pendulum?

P. 595:What major political transformation was taking place in England through the time of Newton's work on gravity?

P. 599:Does the gravity pull of planets and their moons on each other render the solar system unstable?

P. 600:In the Age of Reason, what was Laplace's view of the ultimate power of the human mind?

How was Newton a prime force in the arrival of the Age of Reason?

What impact did the Age of Reason have on political, economic and social thinkers?

Give specific examples for each.

P. 601:Reflect and comment on the contrast between Bondi's and Laplace's positions (p. 599).

P. 602:What were the weaknesses of Newton's theory of gravity that left him and some of his contemporaries with the idea that something was missing?

P. 603:What aspect about the orbit of Mercury remained troubling for the complete gravitational account of the solar system worked out by Newton's successors?

How did small deviations in the predicted orbits of the outer planets lead to new predictions?

What were these predictions and how were these verified?

P. 604:What is the conceptual difference between inertial mass and gravitational mass according to Newton's physics?

Why does a person become weightless in a free-falling elevator?

Why does an astronaut in a satellite orbiting earth become weightless?

P. 605:Why is the satellite also in free fall motion even though it does not fall into the earth?

P. 606:When the floor of an elevator accelerates up toward the ceiling of the elevator, does the "observed acceleration" of an apple released inside the elevator depend on the mass of the apple?

How did Einstein link gravity to motion, and thereby gravitational mass to inertial mass?

P. 608:What new prediction did Einstein make from his equivalence principle?

Compare the magnitude of the deflection angle prediction for the bending of starlight past the sun with the angle of parallax to the nearest star.

P. 609:What would happen to your heart-beat as you approach the massive sun? What would happen to the frequency and wavelength of radio-waves?

What happens to the value of the ratio of circumference of a disk to the diameter of the disk as the disk accelerates rapidly?

P. 610:What is the effect of gravity on the familiar Euclidean properties of space?

P. 611:Without access to the 4th spatial dimension how do you determine if you live in a curved space?

P. 612: How does Einstein's General Relativity dismiss the need for a gravity force?

How do space and time participate in dynamics?

P. 613: What was Einstein's first measure of success for his theory of general relativity?

P. 614: With the arrival of general relativity, did Newton's work become useless?

How have lunar and solar eclipses been key to astronomical advances throughout history?

P. 617: What were the consequences of Einstein's unification of laws of nature for all inertial systems (vessels moving at constant velocity relative to one another)?

What were the consequences of Einstein's unification of laws of nature for all accelerated systems?

Math Based Topics

Forces

Free Fall Motion

Now consider a case where we know the motion. Using the second law, we determine the force. Suppose there is a rock of mass m that is in free fall motion downward. The familiar acceleration is g. Therefore the force that must be acting on the rock is

$$F = ma = mg$$

How does this force arise? The earth attracts the body with force equal to mg. The earth exerts the force of gravity attracting the falling body to the center of the earth. Any object on the surface of the earth experiences a downward force of magnitude (mg). This attractive force is what we have up to now been calling the weight of the body. It is quite distinct from mass.

Example

$$\text{Weight} = \text{Force} = \text{mass} \times \text{acceleration} = \text{mass} \times g$$

$$\text{Weight} = 60\text{kg} \times 9.8\ \frac{\text{m}}{\text{sec}^2} = 588\,\text{N}$$

We also use the units 60 kg wt to refer to weight.

Friction

In describing ideal motion under the guidance of Galileo, we first ignored friction. With the help of Newton's second law we now can add in the effect of friction on the motion. In a later section we will show how to calculate the effect of air resistance, using the ideas of impetus and momentum.

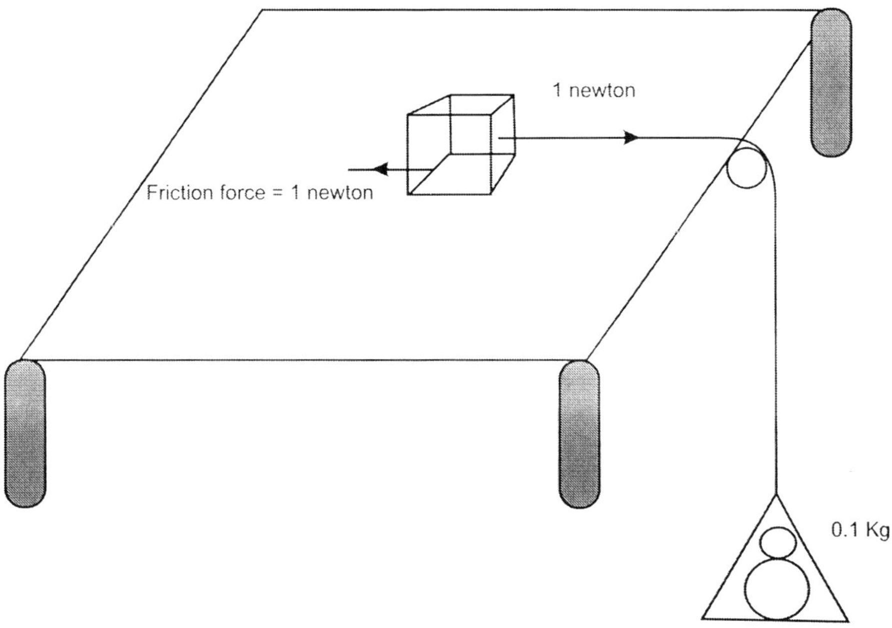

Fig. 11.1 A block at rest on a table is pulled by a string attached to a pan upon which weights are placed so that the object can start to move. If 0.1 kg is needed to overcome friction and start moving the block, the force of friction is measured to be 0.1 kg. Once we know the magnitude of the friction, we can take it into account to determine motion.

In Fig. 11.1, in order for the block (mass = 1 kg) to move with *uniform velocity* a force must be continuously applied to overcome the force of friction between the block and the table, for example, by adding a weight to the pan. The weight will produce a tension in the string. The string therefore exerts a force (equal to the tension) on the block. This force overcomes the friction and the block starts to move. Suppose we determined - by experiment - that the weight added to the pan to get the block to just start moving is 0.1 kg. The force of friction is therefore equal to the weight of 0.1 kG. We do not know much about the detailed nature of the friction. Of course, we can imagine that friction arises from the interactions between the particles of the block and the particles of the surface.

The net force on the block, which is the vector sum of the tension in the string and the friction, will still be zero, since the tension in the string is just sufficient to overcome the force of friction. The block will move, but with uniform velocity. Note that the body is still in equilibrium, since there is no unbalanced force. The acceleration is still zero.

Force of friction = weight in pan = force of attraction of earth on weight = mg

mass = 0.1 Kg

$$g = 9.8\ \frac{\text{m}}{\text{sec}^2}$$

$$\text{weight in pan} = 0.1\ \text{kg} \times 9.8\frac{\text{m}}{\text{sec}^2} = 0.98\ \text{kg}\frac{\text{m}}{\text{sec}^2} = 0.98\ \text{newton} \approx 1\ \text{newton}$$

If we apply a force larger than one newton, there will be an unbalanced force on the block. The block will then start to accelerate. Now we can use the force law to determine the motion. Suppose the mass of the block is 1 kg, and suppose the total force applied is 3 N. The *additional force* is 2 N. Knowing the force, we can calculate the acceleration of the block

$$F = ma$$

$$F = 2\ \text{newton},\ m = 1\text{kg}$$

$$a = \frac{F}{m} = \frac{2\ \text{newton}}{1\ \text{kg}} = 2\frac{\text{m}}{\text{sec}^2}$$

Reaction Force

What about motion perpendicular (⊥) to the plane for a block resting on a flat table? Experience tells us that there is none. In the ⊥ direction, the block is in equilibrium; the acceleration is zero (Fig. 11.2). Therefore, according to Newton's law, the net force on the block in the ⊥ direction must be zero. But we also know that there exists the downward force (mg). We *deduce* that there must be another force acting in the ⊥ direction on the block to cancel the [mg] force. Here we are using Newton's law to derive a force from the observed motion and some of the other forces known. The new force opposes the ⊥ component of mg. We call this the *reaction force*. Since the ⊥ component of the weight is [mg], the reaction force is also

$$|\text{Reaction Force}| = mg$$

Just as in the case of the detailed nature of the friction force, we cannot at this stage further analyze the origin of the reaction force. But we know that it arises as a reaction of the downward weight of the block, due to the detailed interaction between the matter of the block and the matter of the plane upon which it rests. If the block is resting on a horizontal plane, the force of gravity is mg downward, and the reaction force is mg upward. The net force on the block is zero and the block remains in equilibrium (Fig. 11.2). A body resting on a flat surface must therefore have two forces acting upon it: (a) the downward force due to gravity and (b) the upward force of reaction. The two forces are equal in magnitude, but opposite in direction.

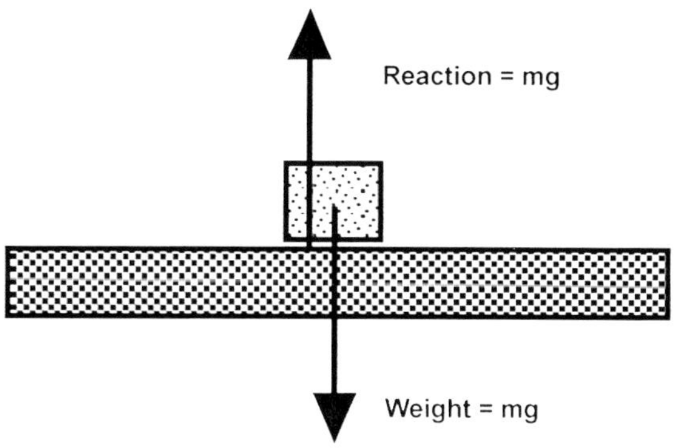

Fig. 11.2 Analysis of forces acting on a block that rests on a horizontal plane.

Circular motion

As the final topic for this chapter, we see how Newton used the motion of the moon to deduce the force acting on the moon. Suppose there is a body of mass m moving in circular motion at constant speed v.

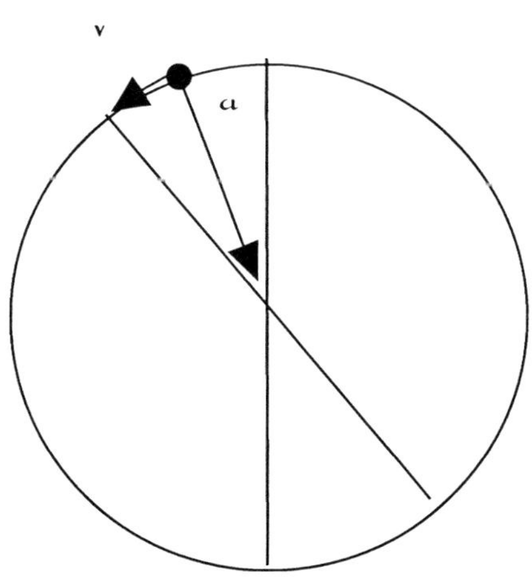

Fig. 11.3 Velocity and acceleration of an object in circular motion.

Since the acceleration $a = \frac{V^2}{r}$ and is pointing toward the center, there must be a force toward the center, and of magnitude $m\frac{V^2}{r}$. This is called the centripetal force which is present for any object in circular motion (Fig. 11.3). In the case of a stone whirling around a string, the centripetal force is supplied by the tension in the string.

Now Newton was faced with a crucial question, which will lead us directly into the next chapter. What supplies the centripetal force for the circular motion of the moon?

Example

A 10 kg mass is attached to a cord which has a maximum breaking strength of 1.8 kN. What is the maximum speed the mass can have if it is whirled around in a horizontal circle with radius = 1 m, and the rope is not to break.

Let V be the maximum speed at which the string breaks. The tension in the string equals the breaking strength. The tension is also equal to the centripetal force = mass x centripetal acceleration.

$$\text{centripetal acceleration} = \frac{V^2}{R} = \frac{V^2}{1} = V^2$$

$$\text{Tension} = \text{massxacceleration} = 10\ V^2$$

$$\text{Tension} = \text{breaking strenght} = 1.8 \times 10^3 = 10V^2$$

$$V^2 = 1.8 \times 10^2$$

$$V = \sqrt{180}$$

$$V = 13.4\ \frac{\text{m}}{\text{sec}}$$

Circular motion

Velocity

$$V = \frac{\text{distance covered}}{\text{time taken}} = \frac{\text{circumference of circle}}{\text{Period}} = \frac{2\pi R}{T}$$

Newton's second law

$$\mathbf{F} = \text{m}\ \mathbf{a}$$

The unit of mass is kilograms (kg), the unit of acceleration is m/sec^2 and the unit for force is therefore kilograms x m/sec^2. This combination is named N, in honor of Newton.

Centripetal Forces

If a body of mass m is in circular motion with a constant speed v, the force required to sustain circular motion is directed toward the center of the circle; the centripetal force is $m\frac{V^2}{r}$

What is the magnitude of the force that the earth exerts on the moon? In the innovative approach we mentioned in Chapter 9, he used the observed motion of the moon to *deduce* the force acting on the moon. Suppose the moon has a mass m and is moving in a circular motion at constant speed V. Since the acceleration is

$\frac{V^2}{r}$ and pointing toward the center, there is a corresponding force of magnitude $m\frac{V^2}{r}$ pointing to the center of the moon's orbit, which is the center of the earth (Fig. 11.4). Newton had shown, independently of Huygens, that the centripetal acceleration of a body in circular motion is V^2/r. He used this formula to make an estimate of the acceleration of the moon, as it goes around the earth in its (nearly) circular orbit. As a first approximation, Newton ignored the fact that the exact orbit is elliptical. His errors were not too large, since the eccentricity of the ellipse is small. He needed the velocity (V) of the moon and the distance (r) from the earth to the moon. The velocity of the moon = $\frac{2\pi r}{T}$ where T is the period of time it takes the moon to complete its orbit around the earth. He used Tycho's detailed observations on the motion of the moon to obtain the exact period of the moon's revolution,

$$T = 29.32 \text{ days} = 2.36\times10^6 \text{ sec}$$

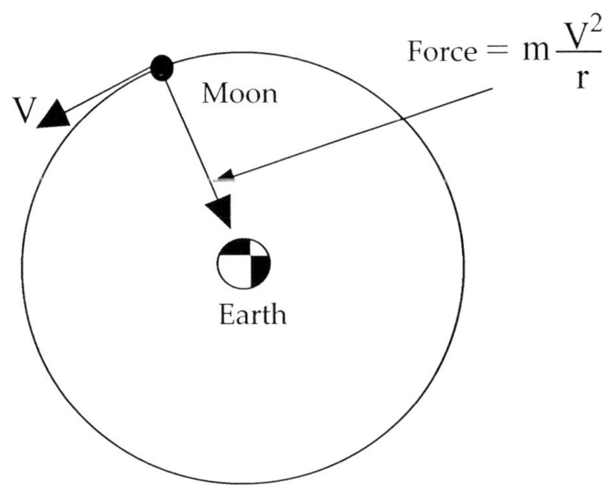

Fig. 11.4: If the moon orbits the earth in a circle with speed V, the centripetal acceleration of the moon is V^2/r, and the corresponding centripetal force on the moon is mV^2/r.

The distance (r) from the earth to the moon is 3.84×10^8 m, 60 times larger than the radius of the earth. Note that we derived an estimate of the distance to the moon in Chapter 4, using measurements carried out by the scientists at Alexandria. Newton had access to better values. With this data, Newton calculated the centripetal acceleration of the moon

$$V = \frac{2\pi R}{T} = \frac{2\pi \times 38.4\times10^7}{2.36\times10^6} = 1021 \text{ m/sec}$$

$$am = \frac{V^2}{R} = \frac{1021^2}{38.4\times10^7} = 0.002715 \text{ m/sec}^2$$

Note how small this acceleration is compared to $g = 9.8$ m/sec^2. Indeed the ratio

$$\frac{9.81}{.00271} = 3609 \approx 3600$$

The acceleration of the moon towards the earth is 3600 times smaller than the acceleration of an earthly object in free fall.

$$a_m = \frac{g}{3600}$$

If a stone on the earth drops 10 meters in one second, the moon drops the same 10 meters in 3600 seconds, i.e., one hour. From his force law, Newton deduced that the force on the moon needed to permit circular motion is

$$F_m = \frac{mg}{3600}$$

where m is the mass of the moon. Compare this force with the hypothetical force on the moon - if the moon were located at the surface of the earth. In this case, the force on the moon would be, F_m=Mg, i.e., 3600 times larger. If the force of gravity that the earth exerts on objects in free fall does indeed extend to the moon, it would be diluted by a factor of 3600. This was the magnitude of the centripetal force on the moon, the answer to the question that Newton raised at the beginning of this inquiry.

Newton was also aware - from Picard's data on the size of the earth - that the radius of the earth is 6.4×10^3 km, which is 60 times smaller than the distance from the earth to the moon. This relation gave Newton a marvelous insight into the nature of the gravitational force. As the moon is 60 times farther from the center of the earth than an apple on the surface of the earth, Newton realized that the force of gravity of the earth when extended to a distant object, such as the moon, *decreases with the square of the distance*. That is to say: the force exerted by the earth on the moon is $60^2 = 3600$ times less than the force exerted by the earth on a falling apple.

Kepler's 2nd Law of Planetary Motion and Newton

First, we need to review an important theorem in geometry. If there are two parallel lines, the area of the triangles of the same base drawn between these two lines are all the same (Fig. 11.5).

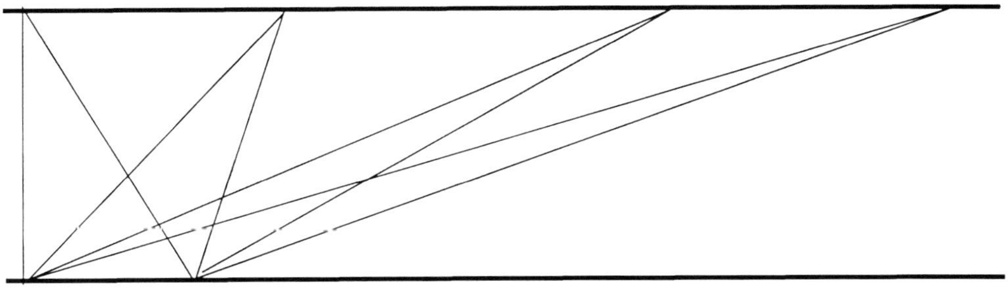

Fig. 11.5: A review of a theorem in geometry. All the triangles drawn between two parallel lines have equal areas.

Below we prove this for the right triangle PAB, and another, more arbitrary triangle QAB, both between two parallel lines (Fig. 11.6):

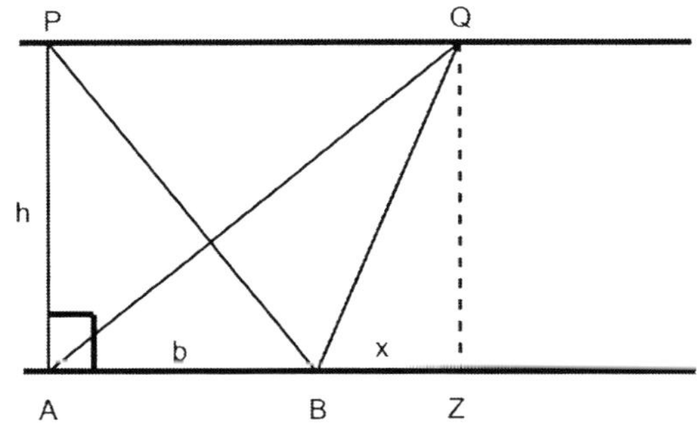

Fig. 11.6: To prove the theorem, we compare the area of an arbitrary triangle AQB to that of the right triangle APB.

Area of the right triangle PAB = 1/2 base × height = 1/2 b h

Area of triangle QAB = Area of QAZ - Area of QBZ

area QAB = 1/2 (b+x) h - 1/2 xh = 1/2 b h = area PAB

Now we can apply this theorem to motion, but first to the case of a body moving with *uniform velocity* V in a straight line (Fig. 11.7). In time equal time intervals, Δt, the body will arrive at the points B,C,D etc.

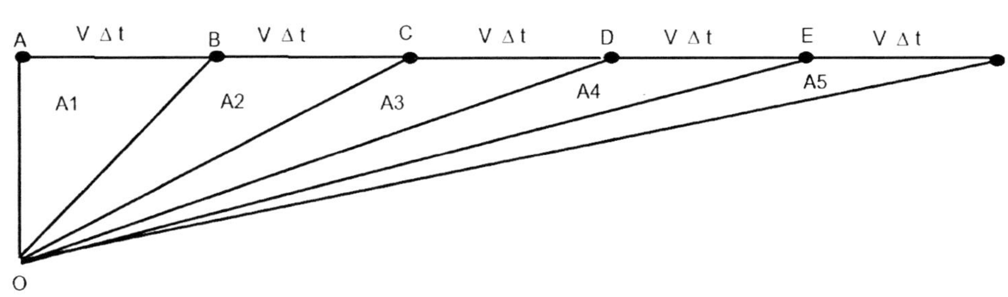

Fig. 11.7: A body moving with uniform velocity in a straight line sweeps out equal areas in equal intervals of time.

The areas A1, A2, A3, A4 etc. are all equal, because the triangles have the same length base (VΔt), and the same height, OA.

Now consider an outside force acting on the body while it is in uniform horizontal motion. Because we want to study planetary orbital motion, we chose the force to be directed toward the center O. This type of force is also known as a *central force*. Suppose the force acts on the body instantaneously when it arrives at B. It causes a change in velocity ΔV. The new velocity is given by vector sum of V and ΔV. Under the influence of the central force, the body is deflected and goes to C′ instead of going to C (Fig. 11.8).

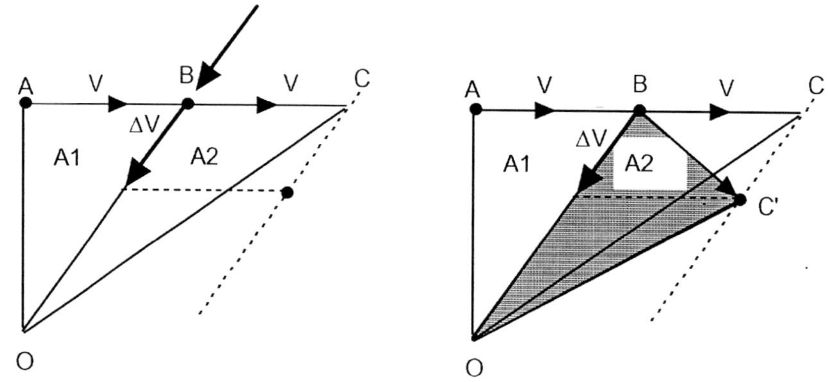

Fig. 11.8: Under the influence of a central force, a uniformly moving body is deflected from its path, and reaches the point C' instead of the point C. The two triangles OBC and OBC' have the same area because of the pair of parallel lines OB and CC'.

The areas of the triangles OBC and OBC′ have the same area because the two triangles are between the same parallel lines (previous theorem). Suppose the force were larger, as for example when a planet is closer to the sun (Fig. 11.9).

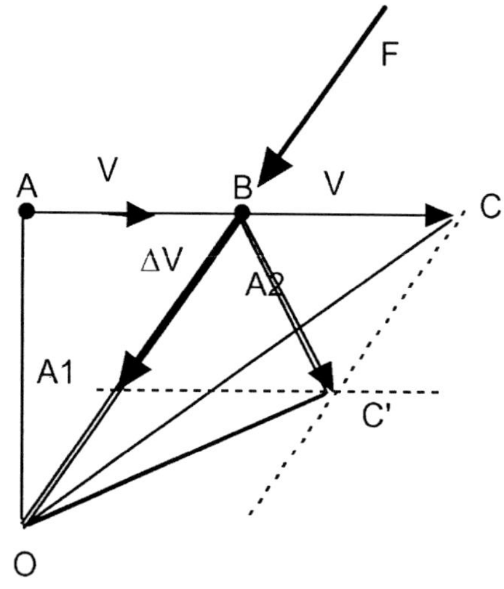

Fig. 11.9: Under the influence of a larger central force, the area of the new triangle OBC' is again the same as the original area OBC.

Again, the area of triangle OBC′ will be the same as OBC. *As long as the force remains directed toward the center,* the area swept out by the planet will be equal in equal time of travel, *regardless of the strength of the force*. Thus Newton proved that Kepler's 2nd Law, the area law, is the direct result of the *central nature of the force* between the planet and the sun. The force acts along the line joining the planets to the sun.

Kepler's Third Law and Newton

Rather than derive the third law, we first show how it leads to the inverse square law of gravitation, and much more. As before, we treat the simpler case of a circular orbit. Kepler showed that for any planet of the sun

$$\frac{R^3}{T^2} = \text{constant}$$

Recall Kepler's stunning realization that the constant is the same for all the planets. If each planet moves in a circular orbit, the acceleration for this circular motion is produced by a force between the sun and the planet. The force is directed toward the center of the circle, i.e., the sun. From Newton's second law of motion, F = ma, we know the force between the planet and the sun

$$F = M_{\text{planet}} \frac{V^2}{R}$$

The planet speed is

$$V = \frac{\text{distance traveled}}{\text{time taken}} = \frac{2\pi R}{T}$$

$$V^2 = \frac{4\pi^2 R^2}{T^2}$$

$$F = M_{\text{planet}} \frac{(4\pi^2 R^2)}{T^2} \frac{1}{R} = M_{planet} \frac{(4\pi^2 R)}{T^2}$$

Now we apply Kepler's 3rd Law:

$$\frac{R^3}{T^2} = \text{constant} = K$$

which is equivalent to

$$T^2 = R^3 \times K'$$

although K′ is the reciprocal of K, it is still a constant

$$F = M_{planet} = \frac{(4\pi^2 R)}{R^3 \times K'}$$

$$F = \frac{4\pi^2}{K'} \frac{M_{planet}}{R^2}$$

Thus, from Kepler's harmonic law, Newton re-established faith in the inverse square law of gravitational attraction. The force that the sun exerts on each planet varies as $1/R^2$.

There is something brand new here. The magnitude of the force depends on the mass of the planet. The constant in Kepler's third law plays a role in determining the magnitude of the force between the sun and the planets. Newton also proved the converse: An inverse square law of attraction leads to Kepler's 3rd Law. We will return to this derivation later to deduce yet another novel aspect of the solar system.

$$F = G\frac{mM}{R^2}$$

Finally, we reach Newton's Law of Gravitation - the mathematical bond between dynamics and astronomy. The law did not come down to us from the mountain top, as did the Ten Commandments. Newton reached it after an arduous climb, assisted by many sudden leaps of creative insight. But Newton was not alone in his quest; it was the culmination of two thousand years of observations, questions and answers, not to mention the many false trails and blind alleys.

G is called the gravitational constant. G should not be confused with g, the acceleration on earth due to gravity, $g=9.8\ m/sec^2$. Later we will see how to determine the value of G. It turns out to be a very small number

$$G = 6.67 \times 10^{-11} \frac{\text{Newton m}^2}{\text{kg}^2}$$

The infinitesimally small value of G indicates that the force of gravity is very weak, unless the masses involved are enormous, such as the mass of the sun or the mass of a star.

Example

Calculate the gravitational force of the moon on the earth and compare it to the force of the sun on the earth. Use the needed mass values from Table 1.

$$F_{moon\text{ - }earth} = \frac{Gm_{moon}m_{earth}}{R_{moon\text{ - }earth}^{2}}$$

$$F_{moon\text{ - }earth} = 6.67 \times 10^{-11} \times \frac{7.4 \times 10^{22} \times 6 \times 10^{24}}{(3.85 \times 10^{8})} = 2 \times 10^{20}\, N$$

$$F_{sun\text{ - }earth} = \frac{Gm_{sun}m_{earth}}{R_{sun\text{ - }earth}}$$

$$F_{sun\text{ - }earth} = 6.67 \times 10^{-11} \times \frac{2 \times 10^{30} \times 6 \times 10^{24}}{(1.5 \times 10^{11})^{2}} = 3.6 \times 10^{22}$$

Pisa Explained

Newton's force law and his law of gravitation, when put together, explain the astonishing result of Galileo's legendary experiment at Pisa. As the mass of an object increases, so does the force of gravitation (Fig. 11.10). Hence the acceleration remains constant. The qualitative account is strikingly simple. Yet it holds the clue to deeper questions about the nature of gravity - which we will touch upon later. Now Newton went a step further. He derived the value of g in terms of the properties of our planet earth, i.e., the mass of the earth and the size of the earth.

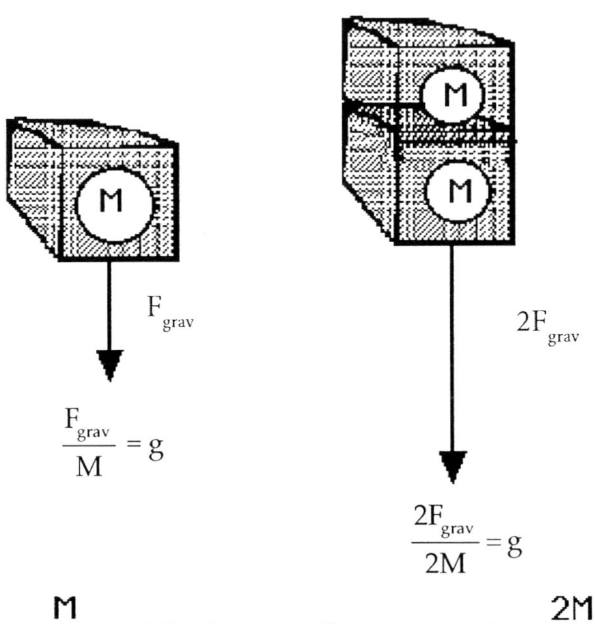

Fig. 11.10 Newton's Law of Gravity and his Force Law together explain why the acceleration of a falling body is independent of its mass.

To determine the acceleration, we start with Newton's Force Law: the acceleration of a body of mass m is given by:

$$a = \frac{F}{m}$$

Now we need to know the force, F. According to the law of universal gravity, the force that the earth exerts on a body of mass (m) is given by:

$$F = \frac{GM_e m}{R_e^2}$$

Remember that it was Newton's long struggle with integral calculus that allowed him to ideally treat the spherically symmetric earth as if it is a single point, with all its mass concentrated at that point. Here is where the mass of the earth (M_e) and the radius of the earth (R_e) come in. Substituting the expression for the force into the expression of the acceleration, followed by some simple algebra

$$a = \frac{\frac{GM_e m}{Re^2}}{m} = \frac{GM_e}{R_e^2} = g$$

The two laws of motion reveal that the *mass (m) of the falling object is not present in the expression for g*! The acceleration due to the earth's gravity is independent of the mass the falling body. When Galileo exposed Aristotle's monumental fallacy, he did not pause too long to wonder why the acceleration is the same for all bodies. He was thoroughly captivated by the universality of free fall motion. Satisfied with properly determining how, he did not jump to the prematurely to the why.

So far the value of g has been considered an arbitrary constant. It could now be related to the physical properties of the earth through the universal gravitational constant, G

$$g = \frac{GM_e}{R_e^2} = G \times (\text{mass of earth})/(\text{radius of earth})^2$$

The Value of G

Of course, Newton did not know the value of G, nor the mass of the earth. All he knew was the radius of the earth - thanks to Eratosthenes and Picard. What Newton actually accomplished was a method to determine the value of G from the already known value of known quantity, g. Recall (Chapter 8) how Huygens had determined a precise value of g from the period of the Galileo's oscillating pendulum. But how to obtain the mass of the earth? Newton approached the problem in terms of the density of the earth - thanks to Archimedes.

$$\text{Density} = \frac{\text{Mass}}{\text{Volume}} = \frac{M}{4/3\pi R_e^3}$$

Here we use the ancient formula for the volume of the spherical earth $V=4/3\pi R^3$. We re-cast the expression for g to *eliminate* the unknown mass of the earth and express g in terms of G.

$$g = \frac{GM_e}{R_e^2} = \frac{GM_e R_e}{R_e^3} \frac{4/3\pi R_e GM_e}{4/3\pi R_e} = 4/3\pi R_e G_x$$

But Newton did not know the density of the earth either. So he made an educated guess:

> "...the common matter of our earth on the surface...is about twice as heavy as water,...in mines, (matter) is found about three or four, or even five times more heavy, it is probable that the quantity of the whole matter of the earth may be five or six times greater than if it all consisted of water..."

Accordingly, Newton took the average density of the earth to be 5.5 times the density of water.

$$\text{Density} = 5.5\frac{\text{gm}}{\text{cc}} = \frac{5.5\times10^{-3}\text{kg}}{10^{-6}\text{m}^3} = 5.5\times10^3\frac{\text{kg}}{\text{m}^3}$$

It proved to be an excellent order of magnitude guess! Scientists often have to make order of magnitude guesses in the face of huge unknowns. Newton was extraordinary lucky in his guess! Finally Newton was ready to calculate a value for G using the known value of $g = 9.8\ \text{m/sec}^2$.

$$g = \frac{4}{3}\pi R_e G \times \text{Density}$$

$$G = g\frac{3}{4\pi R_e} \times \frac{1}{\text{Density}}$$

$$G = \frac{3}{4\pi}9.8\left(\frac{\text{m}}{\text{sec}^2}\right)\frac{1}{6368\times10^3(\text{m})\times5.5\times10^3\left(\frac{\text{kg}}{\text{m}^3}\right)}$$

$$G = 6.67\times10^{-11}\left(\frac{\text{m}}{\text{sec}^2}\right)\frac{1}{(\text{m})\times\frac{(\text{kg})}{\text{m}^3}}$$

$$G = 6.67\times10^{-11}\frac{\text{newtons m}^2}{\text{kg}^2}\text{ in MKS units}$$

A generation later, Cavendish devised a method to measure the gravitational force constant, G, and confirmed that the average specific gravity of the earth is indeed close to Newton's guess of 5.5!

Mass of the Sun

Newton's Laws of motion and the Universal Law of Gravitation lead directly to Kepler's third law:

$$\frac{R^3}{T^2} = \text{Constant}$$

where R is the average radius of the planet's orbit around the sun, and that T is the period of the orbit. But there is much more! We show how Kepler's constant is related to a key physical property of the solar system, *the mass of the sun*. For simplicity of the mathematical treatment, we assume a circular orbit, although with calculus and analytical geometry the treatment can be generalized to elliptical orbits. From Newton's force law, the force on the planet is

$$F_{planet} = M_{planet}\ a_{planet}$$

Here a_{planet} is centripetal acceleration

$$a_{planet} = \frac{V_{planet}^{\ 2}}{R} = \frac{\left(\frac{2\pi R}{T}\right)^2}{R} = \frac{4\pi^2 R}{T^2}$$

Substitute the acceleration into Newton's Second Law

$$F_{planet} = \frac{M_{planet}\ 4\ \pi^2 R}{T^2}$$

Now we need to derive the force. The force on the planet originates from the sun, it is the force of gravity between the planet and sun

$$F_{planet} = G\frac{M_{planet}\ M_{sun}}{R^2}$$

Finally, we are ready to equate F_{planet} from F = ma with F_{planet} from the law of gravity. The combination of the two laws reveals a new result.

$$F_{planet} = \frac{M_{planet} 4\pi^2 R}{T^2} = \frac{GM_{planet} M_{sun}}{R^2}$$

Rearranging

$$\frac{R^3}{T^2} = \frac{GM_{sun}}{4\pi^2}$$

Newton's laws lead directly to Kepler's third law. What is spectacularly new is the mathematical expression for Kepler's constant in terms of the mass of the sun. The radius and period of a planet are intimately connected with the mass of the central body, the sun. The expression involves G, a primary property of the universe, as well as the essential aspect of spherical geometry, the value of π. The legacy of Alexandrian mathematics lives on.

$$\text{Kepler's Constant} = \frac{GM_{sun}}{4\pi^2}$$

A more accurate derivation based on a more general elliptical orbit, not the special case of a circular orbit, gives

$$\text{Kepler's Constant} = \frac{G(M_{sun} + m_{planet})}{4\pi^2}$$

Therefore Kepler's constant is not exactly the same for all the planets, as we claimed earlier. However, since the mass of a planet is much smaller than the mass of the sun, we can continue to use the approximate relationship for Kepler's constant which involves only the mass of the sun, and ignore the mass of the planet. Through the law of gravitation, Newton extended our knowledge of the solar system well beyond a measure of distances. It was now possible to determine the mass of the sun and the planets.

Example

The value of Kepler's constant can be calculated from the radius and the period of the orbit of any one of the known planets. We discussed how astronomers determined the absolute values of the planetary orbits. Their measurements showed that the radius of the earth's orbit around the sun is

$$R = 1.5 \times 10^{11} \text{m}$$

Knowing that the period of the earth's orbit around the sun is one year

$$T = 1 \text{ year} = 31.6 \times 10^6 \text{ sec}$$

$$\text{Kepler's Constant} = 3.4 \times 10^{18} \frac{\text{m}^3}{\text{sec}^2}$$

Example

From Kepler's constant and the value of G, determine the mass of the sun.

$$M_{sun} = \frac{4\pi^2 R^3}{GT^2} = 2 \times 10^{30} \text{kg}$$

The Mass of the Earth

Just as we used the planetary orbits to determine the mass of the central sun, we can use the lunar orbit to determine the mass of the central earth. Consider the earth-moon system as a Keplerian "planetary" system. Applying the same principles as we did to the sun-earth system, we can derive

$$\frac{R^3}{T^2} = \frac{GM_{earth}}{4\pi^2}$$

Here R and T stand for the radius and period of the moon's orbit around the earth. It is very important to note here that Kepler's constant for the earth-moon system is *not* the same as Kepler's constant for the planet-sun system, because now the constant involves the mass of the earth.

Example

Knowing that

$R = 38.4 \times 10^7$ meter = radius of moon's orbit around earth.

$T = 29.32$ day $= 2.36 \times 10^6$ sec = Period of moon's orbit.

Determine the mass of the earth.

$$M_{earth} = \frac{4\pi^2}{G}\frac{R^3}{T^2}$$

Mass of earth $= 6.01 \times 10^{24}$ kg.

The earth is nearly a million times less massive than the sun.

We can similarly determine the mass of Jupiter, by treating the Jupiter-satellite system as a Keplerian system. But we need to know the period of Jupiter's moons, which Galileo already measured with his first telescope. We also need the distance between the moons and Jupiter, which were determined by later astronomers at the Paris Academy - with powerful telescopes. We can repeat the process for Saturn, and find the mass of Saturn, if we know the orbital properties about any of the several moons of Saturn. Finding the moons of the planets and determining their orbital properties leads to a determination of the mass of the planets.

Example: The Geo Synchronous Satellite

Calculate the radius of the orbit of a satellite that revolves around the earth so that its period is exactly 24 hours. Such a satellite would revolve around the earth at exactly the same rate at the earth rotates. It would therefore stay exactly above the same spot all the time. The orbit is called *geosynchronous*. Such a satellite is useful for communications, and is called a syncom satellite. A few geosynchronous satellites placed above the earth would allow radio signals to be transmitted from any point on the earth to any other point.

To calculate the radius of the geosynchronous orbit we use Kepler's third law and apply it to the earth-satellite systems. The moon is a satellite of the earth. In the earth-moon system, R = orbital radius of moon, T = period of moon

$$\frac{R^3}{T^2} = \frac{GM_{earth}}{4\pi^2} = 1.01 \times 10^{13} \frac{m^3}{sec^2}$$

Exactly the same law and the same constant applies to any earth satellite. If T = 24 hours = 8.6×10^4 sec

$$R^3 = 2.4 \times 10^{23}$$

$$R = 4.2 \times 10^7 \text{ m}$$

A satellite orbiting 35.6 million m above the earth would be geosynchronous.

Chapter Questions

1. What is the definition of force?
2. What are the units of force? Give these in terms of m, kg and s.
3. What is the value of the force acting on a body moving with uniform velocity?
4. Discuss *two separate* clues that Newton followed to reach the inverse square law for the gravitational force.
5. List two separate reasons why the value of g on earth is different at the equator than at the poles. What is the consequence of this effect for a pendulum clock?
6. Discuss the connections between the moon and the tides. How does the Law of Gravity provide an explanation for these connections? Which observation confirms the idea that the Moon is primarily responsible for the tides?
7. Why does the moon play a bigger role in the formation of tides than the sun?
8. If a lifeguard at a beach tells you that the high tide is at 9 am, approximately when should you expect the high tide on the next morning?
9. Compare the mass of an astronaut on the Moon with his *mass* on Earth.
10. Beyond the solar system, there are stars that are 10 times as massive as our Sun. What value of G should be used to calculate the gravitational attraction of these stars?
11. An astronaut in the orbiting space station Mir decides to conduct Galileo's experiment and lets go of a feather and a rock that she is holding in each of her hands. Which will reach her feet first?
12. What experiment did Newton propose to prove the rotation of the earth? Draw diagram(s) to describe his ideas.
13. Discuss how Newton's Law of Gravity and his second law of motion explain Galileo's experiment at Pisa.
14. What is the physical evidence for the bulging shape of the earth? Is there any relationship to the flattening at the poles. Discuss the impact of the earth's shape on its motion.
15. What are comets? Discuss the roles played by Newton and Halley in understanding their paths.
16. What were some of the difficulties that Newton's contemporaries had with the concept of universal gravitation?

17. Discuss some of the physics discoveries made at the observatories funded by English and the French kings.

18. Discuss how accurate astronomical measurements were crucial to Newton's success in arriving at the law of gravitation.

19. What new evidence was uncovered between the times of Galileo and of Newton to support the idea of a moving earth?

20. Do Newton's laws of motion and his law of universal gravitation give any information about the cause of gravitational force?

21. NASA wishes to determine an accurate value for the mass of the earth. They propose to do this by sending a satellite into earth orbit. What physical quantities should they be prepared to measure about the satellite, and how would they then use these quantities to achieve the mission goal?

22. Discuss the symmetry and/or unity aspects of Newton's first, second and third laws of motion.

23. Discuss two examples from this chapter of how precision in measurement plays a crucial role in the advancement of science.

24. In Aristotle's view, the laws of motion for water are different from the laws of motion for the substance, earth. In Newton's views, what property distinguishes the motion of objects. Discuss this property.

25. Newton concluded that the shape of the earth is not an exact sphere but that it has a bulge at the equator relative to the pole. What are the effects of this deviation from spherical symmetry?

26. According to the law of universal gravity, there should be a force of attraction between two students. Why do we not observe that the students move toward each other (as the apple moves towards the earth), or why do the students not orbit each other (as the planets orbit the sun)?

27. What is a geosynchronous satellite?

28. An astronaut in the orbiting space station Mir decides to conduct Galileo's experiment and lets go of a feather and a rock that she is holding in each of her hands. Which will reach her feet first?

29. What is the difference between the meaning of inertial mass and gravitational mass according to Newton?

30. What is the difference between the meaning of inertial mass and gravitational mass according to Einstein?

31. What consequence did Newton derive from the assumed equivalence of inertial mass and gravitational mass?

32. What consequence did Einstein derive from the equivalence of inertial mass and gravitational mass?

33. An astronaut in a space capsule full of air at one atmosphere releases a feather and a rock to perform the classical experiment of Galileo. Describe the motion of the two objects.

34. An astronaut takes a bathroom scale on her expedition. She weighs 100 pounds on earth. What does the scale show her weight to be when she stands on the scale while orbiting the earth in the space shuttle?

35. What role does general relativity play in the accuracy of a GPS system?

Math Based Questions

1. What is the value of the force acting on a body moving with uniform velocity?

2. If a body is in equilibrium (no acceleration) what does the sum of all the forces on the body add up to? Explain your result.

3. A 0.25 kg ball is attached to the end of a 0.5 m string and revolved in a horizontal circle at 2.0 m/sec. What net force in newtons is needed to keep the ball in its circular path?

4. A 1500 kg car accelerates uniformly from 44 km/hr to 80 km/hr in 10 sec. Using Galileo's laws of motion, calculate the acceleration in m/sec^2. Then calculate the force exerted on the car in newtons.

5. A 100 kg woman is riding a 20 kg bicycle. She accelerates at 1.5 m/sec^2. How much force (in newtons) does she have to apply with her muscles?

6. A body mass 10 kg is moving with a uniform speed of 10 m/sec in a circle. Calculate the centripetal force. Suppose you double the speed of circular motion. Without re-doing the whole calculation, state the new force needed. Suppose you decrease the radius to one-half its original value of question above, and keep the speed the same. Now what will be the force without re-doing all the calculation.

7. A 1 kg mass is moved horizontally with a force of 10 N over a distance of 10 m on a surface where there is no friction. Calculate the acceleration and the final velocity.

8. Make three columns labeled: a) vector, b) scalar and c) units. Then place the following physical quantities in the appropriate column 1 or 2 and give the units of the quantity in column 3:

   time, mass, length, force, velocity, acceleration, momentum, displacement, distance traveled.

9. If the dimensions of length, time and mass are L, T and M, respectively, then what are the dimensions of G?

10. Calculate the gravitational force between two persons, each with mass = 100 kg and standing one meter apart. Compare the force with the gravitational attraction of the earth.

11. Show that the force of gravity that Jupiter exerts on earth is very small compared to the force exerted by the sun.

12. The value of g will different at the pole than at the equator. We can estimate the difference in g from

$$g = \frac{GM_{earth}}{R^2}$$

If the radius of the earth near the equator is Req = 6401 km, and the radius of the earth near the pole is R pole = 6380 km, estimate the per cent difference in the weight of a body at the pole and at the equator using the assumption that all the mass is concentrated at the center.

13. Compute the centripetal acceleration of the earth toward the sun in m/sec^2 (assume a circular orbit). From the mass of the earth, compute the force that the sun exerts on the earth. Now compute the gravitational force that the sun exerts on the earth using Newton's law of gravity, the mass of the earth, the mass of the sun and the distance between them. Do your two answers agree? If there is a discrepancy, what is the reason?

14. Determine the ratio of the mass of the sun to the mass of the earth. From the radius of the sun and the mass of the sun, determine the density of the sun. Why do you think that the density of the sun is less than that of the earth?

15. Find the value of g (acceleration downward due to the force of gravity) on the moon. Express your answer as a fraction of g on earth.

16. A pendulum has a period of one second. If it is transported to the moon (hint: where the value of g is different from g on earth), what is the period of the same pendulum?

17. Calculate the gravitational force between two persons, each with mass = 100 kg and standing one meter apart. Compare the force with the gravitational attraction of the earth.

18. From the motion of any of the moons of Jupiter, calculate the mass of Jupiter. Repeat your calculation using the data for a different moon. Are you answers the same? Why?

17. The acceleration due to gravity on the surface of Mars is 3.7 m/sec^2. If the planet's diameter is 6.8×10^6 m, determine the mass of Mars and compare it to that of the earth.

19. Saturn has a moon which orbits the planet with a period of 15.91 days. If the radius of the orbit is $12^2\times10^9$ m, calculate the mass of Saturn. Make sure to use the MKS system of units.

20. How high above the surface of the earth do you have to go so that the acceleration due to gravity decreases from 9.8 to 1 m/sec^2?

21. Determine the acceleration of the moon toward the earth. Determine the acceleration of the earth toward the moon

22. Suppose the earth was twice as large in size but its mass was the same. What would happen to the value of g?

23. Determine where between the sun and the earth a satellite must be positioned so that the force of gravitational attraction due to the sun is exactly equal (and in opposite directions of course). Express the distance in terms of the radius of the earth.

24. Using the values of g, calculate the weights of a one kg mass object at the pole and at the equator.

25. Calculate the force of gravity between two stars that are as massive as the sun. Use the distance from the earth to the nearest star as the separation between the two stars. Compare this to the force between the earth and the sun.

26. In one of the worked out examples we compare the gravitational force of the sun on the earth with force of the moon on the earth. Now consider the "tidal perturbation," and replace R^2 by R^3. Compare the tidal perturbations of the sun and the moon on the earth.

27. Suppose there is a meteor high above the Earth, so that its height above the surface is one Earth radius. What is the value of its acceleration toward the Earth?

28. Compute the centripetal acceleration of the earth toward the sun in m/sec^2 (assume a circular orbit). From the mass of the earth, compute the force that the sun exerts on the earth. Now compute the gravitational force that the sun exerts on the earth using Newton's law of gravity, the mass of the earth, the mass of the sun and the distance between them. Do your two answers agree? If there is a discrepancy, what is the reason?

Sample Quiz Questions

1. Mass is a measure of

   a. How much force is needed to stop a moving object.

   b. How much force is needed to accelerate an object.

   c. How much force is needed to change the direction of a moving object.

   d. All of the above.

   e. (a) and (b).

2. Compute the centripetal acceleration of the earth toward the sun in m/sec2 (assume a circular orbit). Earth-sun distance = 1.5×10^{11} m

   a. 3×10^4 m/sec^2 b. 21.4 m/sec^2

   c. 6×10^{-3} m/sec^2 d. 3.43 m/sec^2

3. A 100 kg woman is riding a 20 kg bicycle. She accelerates at 1.5 m/sec2. How much force (in newtons) does she have to apply with her muscles?

   a. 180 newton b. 1176 newton

   c. 120 kg d. 80 newton

   e. 150 newton

4. Suppose there is a meteor with height above the surface of earth equal to one Earth radius. What is the value of the acceleration toward the Earth?

   a. 2.45 m/sec^2 b. 9.8 m/sec^2

   c. zero d. about 5 m/sec^2

   e. none of the above

5. If the centripetal acceleration of the Moon toward the earth is 0.0027 m/sec^2, determine the force by which Earth attracts the Moon.

   a. 10^{20} newton b. 2×10^{20} newton

   c. 2×10^{14} newton d. 7.25×10^{23} newton

6. Saturn has a moon which orbits the planet with a period of 15.91 earth-days. If the radius of the orbit is 1.22×10^9 m, calculate the mass of Saturn. Make sure to use the MKS system of units.

   a. 5.7×10^{24} kg b. 5.7×10^{26} kg

   c. 6×10^{24} kg d. 5.7×10^{24} g

   e. 5.7×10^{27} kg

7. Calculate the value of g in m/sec^2 on the moon.

a. 1.63 b. 9.8

c. 3.3 d. 4.9

8. If the lifeguard at a beach tells you that the high tide is at 9 am, approximately when should you expect the high tide on the next morning?

a. 9 am b. 8 am

c. 10 am d. 11 am

9. Newton concluded that the shape of the earth is not an exact sphere but that it has a bulge at the equator relative to the pole. What is the effect of this deviation from spherical symmetry?

a. The distance along the earth for one degree of latitude at the pole is larger than the corresponding distance for one degree of latitude near the equator.

b. Over centuries the sun will rise in different constellations during the first day of spring.

c. The axis of rotation of the earth precesses slowly.

d. All of the above.

e. b. and c.

10. Why does the moon play a bigger role in the formation of tides than the sun?

a. Because the force of gravity of the moon on the earth is larger than the force of gravity of the sun on the earth.

b. Because the moon rises about one hour later every night.

c. Because the change in the moon's gravitational force across the earth's diameter is much larger than the change of the sun's force across the earth's diameter.

d. Because the moon shows the same face to the earth.

e. None of the above.

Answers to Sample Quiz for Chapter 11

1. d
2. c
3. a
4. a
5. b
6. b
7. a
8. c
9. d
10. c

Homework Questions for Chapter 11 (with sample answers)

1. What is the value of the force acting on a body moving with uniform velocity?

2. List two separate reasons why the value of g on earth is different at the equator than at the poles. What is the consequence of this effect for a pendulum clock?

3. If a lifeguard at a beach tells you that the high tide is at 9 am, approximately when should you expect the high tide on the next morning?

4. Compare the mass of an astronaut on the Moon with his mass on Earth.

5. Beyond the solar system, there are stars that are 10 times as massive as our Sun. What value of G should be used to calculate the gravitational attraction of these stars?

6. A 0.25 kg ball is attached to the end of a 0.5 m string and revolved in a horizontal circle at 2.0 m/sec. What net force in newtons is needed to keep the ball in its circular path?

   We want to use F = ma to find the force needed to keep the ball moving in a circle. We know the mass, so we need to find the acceleration. Using our circular motion equation:

$$a = \frac{v^2}{R}$$
$$a = \frac{(2.0m/s)^2}{0.5m}$$
$$a = 8m/s^2$$

   Using this value for acceleration, F = ma gives us F= 2N.

7. A 1500 kg car accelerates uniformly from 44 km/hr to 80 km/hr in 10 sec. Using Galileo's laws of motion, calculate the acceleration in m/sec^2. Then calculate the force exerted on the car in newtons.

8. Compute the centripetal acceleration of the earth toward the sun in m/sec^2 (assume a circular orbit). From the mass of the earth, compute the force that the sun exerts on the earth. Now compute the gravitational force that the sun exerts on the earth using Newton's law of gravity, the mass of the earth, the mass of the sun and the distance between them. Do your two answers agree? If there is a discrepancy, what is the reason?

9. Find the value of g (acceleration downward due to the force of gravity) on the moon. Express your answer as a fraction of g on earth

10. Calculate the gravitational force between two persons, each with mass = 100 kg and standing one meter apart. Compare the force with the gravitational attraction of the earth.

Calculate the gravitational force between two persons, each with mass=100kg and standing one meter apart. Compare the force with the gravitational attraction of the earth.

Again, we're using good old:

$$F = \frac{GMm}{R^2}$$

With M = m = 100kg, and R = 1m, we get $F_{person} = 6.67 \times 10^{-7}$N, a very small force. To get the Force between a person and the earth, we can use this approach with M as the mass of the earth, and R as the Earth's radius (m is still 100kg). This gives $F_{earth} = 980$N. The ratio between these two numbers is 1.5×10^{9}. Earth's pull on either person is 1.5 Billion(!) times bigger than the pull between the two people.

11. Saturn has a moon which orbits the planet with a period of 15.91 days. If the radius of the orbit is 1.22 × 10^9m, calculate the mass of Saturn. Make sure to use the MKS system of units.

CHAPTER 12

Reading Questions

P. 619: What incredible prediction did Einstein's general theory make about the nature of the universe?

Why was the prediction incredible to Einstein?

Why did Einstein abandon his own prediction?

What did he do about it?

P. 620: What is the Doppler effect for sound? How is it used by police?

What is the Doppler effect for light? How is it used by astronomers?

What does a redshift in a spectral line indicate about the motion of a star?

What does a blueshift in a spectral line indicate about the motion of a star?

What did Slipher notice about the redshift of dimmer nebulae?

P. 621: How did Hubble determine distances to extra far away galaxies that had no Cepheid variables?

What evidence did Hubble gather to show that the universe must be expanding?

Are we at the stationary center of the universe's expansion?

What does the expansion of the universe mean for the dynamics of space?

P. 622: Besides the expansion of the universe, what other crucial information does Hubble's law provide?

How did Baade improve Hubble's law? What was the impact?

P. 623: How does Hubble's expanding universe lead to a model for the origin of the universe?

P. 625: Why could the elements heavier than helium not form during the big-bang event, or soon thereafter?

P. 626: How do we know that only hydrogen, helium and trace quantities of lithium formed during the big bang?

At what stage after the big bang did light escape?

P. 627: By how much has the wavelength of the first light shifted since the light escaped after the big bang?

What is the corresponding temperature of a body that emits radiation at such a wavelength?

P. 628: How was after-glow from the big-bang first detected?

P. 629: Give three items of evidence that support the occurrence of the big bang?

How do we know that the electromagnetic and weak forces were the same at some time after the big bang?

P. 630: How is the microcosmos of quarks and leptons related to the macrocosmos of the big bang?